IMAGES OF WAR

HITLER'S VENGEANCE WEAPONS

V1, V2 & V3 LAUNCH SITES

RARE PHOTOGRAPHS FROM WARTIME ARCHIVES

Ian Baxter

Pen & Sword
MILITARY

First published in Great Britain in 2026 by
PEN & SWORD MILITARY
an imprint of Pen & Sword Books Ltd
Yorkshire – Philadelphia

ISBN 978-1-03612-880-7

A CIP catalogue record for this book is available from the British Library.

Typeset by Concept, Huddersfield, West Yorkshire, HD4 5JL.
Printed and bound in England by CPI Group (UK) Ltd, Croydon, CR0 4YY.

FSC www.fsc.org MIX Paper | Supporting responsible forestry FSC® C013604

The Publisher's authorised representative in the EU for product safety is Authorised Rep Compliance Ltd, Ground Floor, 71 Lower Baggot Street, Dublin D02 P593, Ireland – www.arccompliance.com

For a complete list of Pen & Sword titles please contact
PEN & SWORD BOOKS LTD
47 Church Street, Barnsley, South Yorkshire, S70 2AS, England
E-mail: enquiries@pen-and-sword.co.uk
Website: www.pen-and-sword.co.uk
or
PEN & SWORD BOOKS
1950 Lawrence Road, Havertown, PA 19083, USA
E-mail: uspen-and-sword@casematepublishers.com
Website: www.penandswordbooks.com

Contents

About the Author

Ian Baxter is a military historian who specialises in German twentieth-century military history. He has written more than seventy books, including *Poland: The Eighteen-Day Victory March*; *Panzers in North Africa*; *The Waffen-SS Ardennes Offensive*; *The Western Campaign*; *The 12th SS Panzer Division Hitlerjugend*; *Waffen-SS on the Western Front*; *Waffen-SS on the Eastern Front*; *The Red Army at Stalingrad*; *Elite German Forces of World War II*; *Armoured Warfare: German Tanks of World War II*; *Blitzkrieg*; *Panzer Divisions at War*; *German Armoured Vehicles of World War Two*; *Last Two Years of the Waffen-SS at War*; *German Soldier Uniforms and Insignia*; *German Guns of the Third Reich*; *From Retreat to Defeat: The Last Years of the German Army at War 1943–45* and, most recently, *The Sixth Army and the Road to Stalingrad*.

He has written over a hundred articles, including 'Last days of Hitler', 'Wolf's Lair', 'The Story of the V1 and V2 Rocket Programme', 'Secret Aircraft of World War Two', 'Rommel at Tobruk', 'Hitler's War with his Generals', 'Secret British Plans to Assassinate Hitler', 'The SS at Arnhem', 'Hitlerjugend', 'Battle of Caen 1944', 'Gebirgsjäger at War', 'Panzer Crews', 'Hitlerjugend Guerrillas', 'Last Battles in the East', 'The Battle of Berlin' and many more.

He has also reviewed numerous military studies for publication, supplied thousands of photographs and important documents to various publishers and film production companies worldwide, and he lectures to schools, colleges and universities throughout the United Kingdom and the Republic of Ireland.

Chapter One

Rocket Technology

Development of Hitler's 'Retaliation' weapons, although not under that name, had begun in earnest with a multitude of experiments with rocket technology in the early 1930s. The Wehrmacht appointed a 35-year-old Reichswehr captain named Walter Dornberger, who immediately set to work on a rocket prototype called Aggregat 1, or A-1. Whilst he was researching the development of the A-1, he hired a young rocket enthusiast named Wernher von Braun. Between 1930 and 1932, the A-1 was designed and built at the Army Weapons Department at Kummersdorf, south of Berlin. It weighed about 330 lbs, but was not balanced properly and failed to fly. While Dornberger laboured to rectify the A-1 faults, his friend and associate, Wernher von Braun, with an independent team of specialists, developed the A-2. Although it was nothing more than an experimental liquid-fuelled rocket, stabilised by a gyro-mechanism, in 1934 it was capable of being fired to an altitude of 7,000 feet. Because the A-2 was too small for military use, von Braun set to work on a larger, heavier model, the A-3.

The new missile was regarded as Germany's most secret military weapon, and because the authorities were very concerned about prying eyes at Kummersdorf, another site would have to be found for testing the rockets. In 1935, a secluded site was found, situated on the island of Peenemünde, just off the northern coast of Germany, near to the small town of Wolgast on the Baltic Sea – the island was ideal. It was not only densely wooded and sparsely populated, but the area offered a 300-mile long range for test shots. A year later, the Wehrmacht and the Luftwaffe jointly purchased the island. In 1937, planners got to work building a new housing estate for the scientists, engineers and administrators and their families. Soon Peenemünde had all the characteristics of a model village, and it was home to more than 3,000 people.

South of the housing estate there were barracks erected for the soldiers of the research centre and a PoW camp for foreign workers which also included conscripted workers, mainly from Poland. In the southwest corner of the peninsula, a huge factory had been constructed to produce liquid oxygen which was used as rocket fuel. Test stands, wind tunnels, and a massive 30,000-kilowatt power station were also located here. The heart of the rocket research centre, however, was hidden in a forest near the northeastern corner of the peninsula. It was here the experimental, research development laboratories and headquarters office block were situated.

While the Peenemünde rocket centre took shape, development of the A-3 continued. Although the rocket was temperamental, a great deal of information was learned from its design and capabilities. In 1939, a redesigned model was revealed – hot off the drawing board, called the A-4. Despite Hitler's enthusiasm, he was by this stage immersed in war against Europe. Poland had been conquered and in 1940 Holland, Belgium and France capitulated. At this point, the Führer was not convinced the A-4 would have to be built. However, when the Luftwaffe struggled to beat the RAF during the Battle of Britain in September 1940, the A-4 was once again given top priority classification. Test firing once again resumed of the A-4 in March 1942. The 46-foot-long weapon was a 12-ton rocket which could carry a 1-ton warhead 200 miles, and became known as the V-2. The first V-2 launch blew-up. Three months later, a second rocket went out of control after liftoff and crashed in the Baltic Sea. Finally, the most critical test in October 1942 made a successful launch in front of a nervous team of experts, and a number of cameras. Both Dornberger and von Braun breathed a sigh of relief and were hopeful of Hitler's support.

As further tests continued, in July 1943, Dornberger managed to steal a very important meeting with Hitler and Albert Speer, Minister of Armaments and War Production, at Hitler's East Prussian headquarters, Wolfschanze. Apart from being shown some scale models of the V-2 launchers and a few coloured graphs and numerous drawings, they were also shown an impressive colour film of the October launching, with the distinctive black-and-white painted rocket streaming through the sky and splashing down on target in the Baltic Sea.

Hitler was so impressed that he immediately demanded that Donrberger and von Braun build a rocket that would carry 10 tons of explosive. When the scientists replied that it would take nearly five years to develop such a missile, Hitler, not discouraged, ordered Organization Todt, the civil and military engineering organisation of the Reich, to commence at once building mammoth concrete launch bunkers for the V-2s. However, Dornberger stressed his concerns about Allied air-attacks on the proposed V-2 installations, and favoured small, mobile launchers, which would be virtually undetectable from the air. But the Führer's orders prevailed, and the V-2 went into mass production along with a number of concrete launching bunkers.

While the Wehrmacht was developing and experimenting its long-range rockets at Peenemünde, a team of Luftwaffe scientists frantically set to work developing a new weapon that was not only far simpler than the V-2 rocket, but could be manufactured in less time for far less money than the Wehrmacht's complicated guided missiles. Experts began designing a new rocket, and within months Luftwaffe designers had drawn-up a blueprint for a small expendable pilotless aircraft, known officially as FZG 76 – Flakzielgerat (Anti-aircraft Target Device) 76, or better known as the flying bomb or V-1.

For Hitler and Reichsmarshall Herman Göring, Commander-in-Chief of the Luftwaffe, the V-1 offered a means of retaliating against incessant Allied air bombings without risking the lives of badly depleted Luftwaffe crews. With this in mind, they accepted the proposals of Field Marshal Milch of the German Air Ministry, and officially commenced its development on the western peninsula of Peenemünde. It was here that the flying bomb, now code-named Kirschkern (Cheery stone), practically took over all Luftwaffe missile developments on the site.

The design of the V-1 resembled a small aircraft with a stovepipe over its tail and no cockpit. It was about 25 feet in length, with a wingspan of just over 17 feet. The jet engine, which was housed in the 'stovepipe' assembly, was fuelled by standard 80 octane petrol. It carried a 1-ton warhead, the same as the V-2 rocket. The bomb was launched by catapult or launching ramp, which varied to a length of 158 feet overall and was inclined at an angle. Once the rocket was fired, it accelerated a sledge carrying the V-1 along the inclined ramp until it was airborne. As soon as the bomb left the ramp, the leading burnout element of the sledge fell away. In flight, the bomb was on its own, flying along a pre-set gyroscope-controlled course.

Although the V-1 was not accurate at hitting a pre-selected target, it would, however, be accurate enough to hit a target as large as London. For this reason, intensive testing of the pre-production models continued into the early months of 1944, in preparation for a flying bomb assault against England. Göring had become so confident of his 'Wunderwaffe', that he declared during a conference he was expecting initial production of 'Cheery stones' at the rate of about 50,000 a month. Commencement of this unremitting attack on England was set for Christmas 1943. But there was division among senior Luftwaffe commanders on the method of launching sites. Milch favoured a small number of giant bomb-roof bunker installations. General Walter von Axthelm, Ministry of Aviation with the position of General der Flakwaffe (Inspector of Anti-Aircraft Artillery), however, was opposed to the bunkers, outlining that the constructions would be easy prey to enemy aircraft, and bombed before they were completed. He suggested that dozens of small mobile launching sites be erected throughout northern France, as it was doubtful the enemy would be able to destroy them all. In the event that no clear decision was made and both ideas were constructed, it was also planned to air-launch V-1s from Heinkel HeIll's as well.

Altogether there were to be 100 small launching sites erected, and four bombproof bunkers. Work began immediately. To man these sites, Luftwaffe flak units were drafted into special Flakregiments code-named 155 (W) in Tittensorort, which formed on 15 August 1943 under Colonel Max Wachtel. In total, there would be sixteen batteries subordinated to the 5th Flak Division (W).

From the western side of Peenemünde, the launching crews began a series of extensive firing trials. Due to delays caused by insufficient supplies of V-1s for testing, mainly due to the defective quality of its design, flak crew used early experimental models in practice shots, which were fired downrange in rapid succession over the Baltic Sea.

By October 1943, only thirty-eight V-1s were delivered from the manufacturers. The Volkswagen plant at Fallersleben, near Hamburg, had been expected to turn out at least 5,000 a month. But with an endless series of problems constantly afflicting the bomb, technicians were hard pressed to achieve their Christmas deadline. Consequently, the mass-production timetable was put in jeopardy, and following a series of other major technical issues, the deadline was eventually pushed back to spring 1944.

Finally, by the end of March 1944, an operational flying bomb was ready. In April, 1,000 FZG 76s rolled off the assembly line. One month later, another 1,500 were delivered. In June, 2,500 more were produced and ready for action. Now that it appeared there were enough flying bombs stockpiled, Hitler, eager more than ever to wipe London off the face of the map, now sent word through to the Wehrmacht High Command to order Watchtel to prepare his flying bombs for an attack on England for 10 June.

(**Below**) A photograph showing Walter Dornberger and Fritz Todt at the Peenemünde Army Research Centre in 1941.

(**Opposite, above**) Dornberger, Olbricht, Brandt and von Braun at the Peenemünde Army Research Centre.

(**Opposite, below**) General Emil Leeb, Heinrich Lübke, Fritz Todt, and Walter Dornberger at Peenemünde Army Research Centre in the spring 1941.

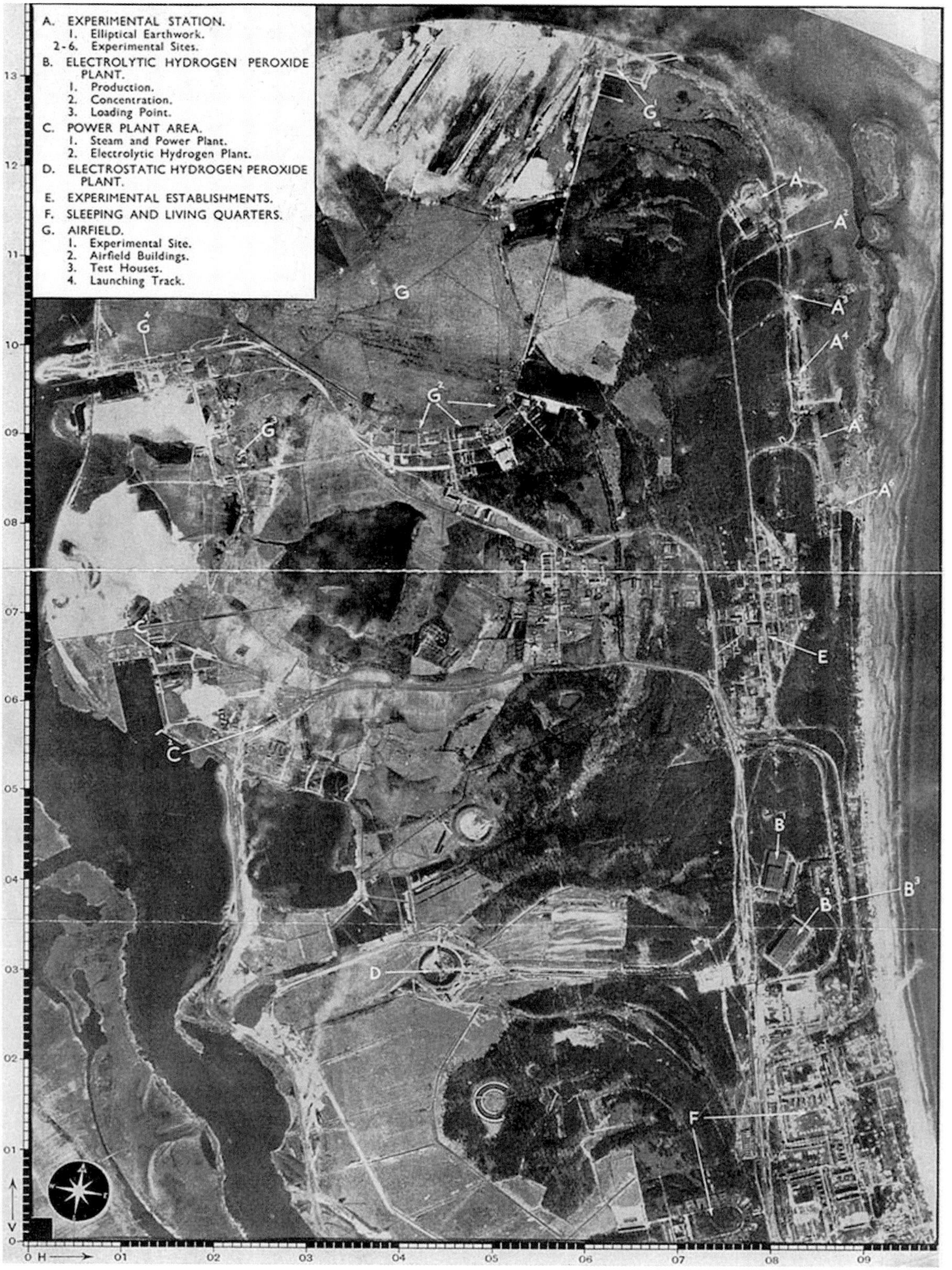

British aerial reconnaissance photograph of Peenemünde taken in April 1943.

A control unit of a V-2 rocket taken at Peenemünde probably in 1942 or 1943.

A photograph taken inside the facility at Peenemünde showing V-2 rocket workshop.

Known as the Meillerwagen (Trailer) which was a complex transport trailer essential for moving and deploying the V-2 rocket for launching. It had an elevating mechanism to raise the rocket vertically for launch.

Four photographs taken at the Peenemünde rocket facility showing a V-2 rocket on a special rail car. V-2 rockets were transported by train as part of their deployment strategy, arriving in launching areas under the cover of darkness for concealment. From the railhead, they were transferred to mobile Meillerwagen trailers for transport and erection at the launch site, where they were then fuelled with liquid oxygen and alcohol before firing. This train transport was vulnerable to Allied attacks, and Allied efforts focused on bombing production facilities and attacking rail transit to disrupt their deployment.

Launching crew can be seen fuelling a V-2's fuel tanks with liquid oxygen whilst on a rail car.

A V-2 rocket tank at Peenemünde rocket facility. The V-2 rocket had two main propellant tanks that made up most of its body: the upper tank stored a mixture of alcohol and water, and the lower tank held liquid oxygen, which was super-cooled. These two tanks provided the fuel for the V-2's liquid-fuelled engine, allowing it to fly to high altitudes and strike targets with its warhead.

Two photographs showing an elevating mechanism which has been used to raise the rocket vertically for launch. Note the adapted service gantry function, which allowed crews to fuel and prepare the rocket after it was erected. These rocket gantries referred to the mobile, 15-tonne, 1942 German Fries/Strabo portable gantry crane used by Wehrmacht maintenance units for tasks such as lifting tank turrets, and also by V-2 crews. These gantries, designed for quick erection, were an important part of the ground support equipment for the V-2 rocket.

At the Peenemünde rocket facility is a modified Sd.Kfz 10 halftrack vehicle at Peenemünde that was used to tow the Meillerwagen A-4/V-2 transport-launcher unit.

An Opel Blitz 3t V-2 rocket T-Stoffwagen tanker truck.

At the Peenemünde rocket facility, Soviet PoWs rest. Soviet and Polish prisoners made up a large portion of the forced labour at Mittelbau-Dora, alongside inmates from France, Belgium, and other occupied countries. These labourers were deemed inferior to Western Europeans by the Nazi racial hierarchy and were assigned the most dangerous and gruelling tasks.

Russian PoWs working on lathes under the supervision of the Wehrmacht at the Peenemünde Research Centre.

Two photographs showing Soviet PoWs working on technical drawings at Peenemünde.

A photograph showing preparations to launch the V-2 at the Peenemünde rocket facility. The process of launching the V-2 began with the weapon being raised via the Meillerwagen lift frame. Once it was vertical, the rocket was suspended above the firing stand, which was raised to touch the rocket fins. Clamping collars were used to connect the rocket to the firing stand, and the Meillerwagen was withdrawn and the rocket was turned a quarter turn for the fuel connections to face the Meillerwagen platforms which were about a metre away, which were lowered into position for standing.

(**Above**) On their launching pads are three V-2 rockets. A separate three-section extendable ladder, the Magirus ladder, was used for areas of an erected rocket not accessible via the Meillerwagen.

(**Opposite**) The first of seven photographs showing the launching of the V-2 at the Peenemünde rocket facility. The launch process for a V-2 rocket involved transporting the missile to a remote site, positioning it vertically on a mobile launching table (Meillerwagen), performing a dry nitrogen purge of the tanks, and then fuelling it with alcohol and liquid oxygen. After final checks, the rocket's internal gyroscope and guidance systems were set, an electrical connection was made, the site was cleared, and the igniter sequence initiated the fuel flow and powerful thrust for liftoff. Pressurized nitrogen forced hydrogen peroxide and sodium permanganate into the steam generator, producing high-pressure steam. This steam powered the turbo pump, forcing fuel into the combustion chamber. The mixture was ignited, the engine reached full thrust, and the electrical connection was severed, allowing the rocket to lift off the launching table. A special launch switch was activated by the motion, and the rocket's gyroscopic guidance system began pitching the vehicle toward its target. The engine burned for approximately 60 to 70 seconds before a timed mechanism cut off the fuel supply, ending powered flight. The rocket continued on an unpowered, parabolic ballistic trajectory, with no further control possible, and re-entered the atmosphere at supersonic speed.

A photograph of Reichsmarshal Hermann Göring who was the head of the Luftfarhtforschungsanstalt Herman Göring (Hermann Göring Aeronautical Research), but Werner von Braun's team at Peenemünde was the one that developed and launched the actual V-2 rockets. Göring also made the decision to launch V-1 attacks, and some accounts mentioned his plans for rocket attacks on New York.

(**Left**) Field Marshal Erhard Milch served as a key member in the development and production of the Luftwaffe. As State Secretary in the Reich Ministry of Aviation and Inspector General of the Luftwaffe, he held immense power over German air armament during the war. Along with his governmental task force, his objective was to increase the production of fighter aircraft, in part by moving the production facilities underground. In cooperation with the SS, the task force played a key role in the exploitation of forced labour for the benefit of the German aircraft industry and the Luftwaffe, including the forced labour to build V-2 rockets underground.

(**Right**) Walther von Axthelm was Inspector of Anti-Aircraft Artillery: He also held the role of General der Flakwaffe (Inspector of Anti-Aircraft Artillery) from 1942 to 1945 in the Ministry of Aviation. It was Axthelm who was opposed to building launching bunkers, outlining that the constructions would be easy prey to enemy aircraft, and bombed before they were completed. He suggested that dozens of small mobile launching sites be erected throughout northern France, as it was doubtful the enemy would be able to destroy them all.

(**Opposite, above**) V-1 flying bombs are being towed out following production. The V-1 was primarily constructed with a welded sheet steel fuselage and plywood wings. It featured an Argus pulsejet engine mounted on top, a small propeller to calculate range, and an autopilot with a gyroscope and air log for guidance. Compressed air powered the internal servos for the rudder and elevators. The missile carried an Amatol warhead and was launched from ramps. Note that the wings have not been attached; these were attached after transportation to their launching site. The wings were constructed from plywood, though some later, longer-range models featured plywood wings on a tubular metal spar.

(**Below**) RAF Reconnaissance photograph showing the aftermath of Allied bombing on Peenemünde. British intelligence had collated information, including data from forced labourers, confirming that Peenemünde was the German Army's secret research centre for the V-weapons. The long-range V-weapons were seen as a grave threat to Great Britain and had to be attacked. The bombing raid was an attempt to destroy the research facilities and kill the scientists, thereby delaying the programme. The Peenemünde raid was the first offensive in the larger strategic bombing campaign known as Operation Crossbow, which targeted all German V-weapon sites. The RAF lost 40 bombers and 215 aircrew members. German casualties comprised of approximately 180 German personnel killed, including two V-2 rocket scientists. Rocket engineer Werner von Braun survived the attack.

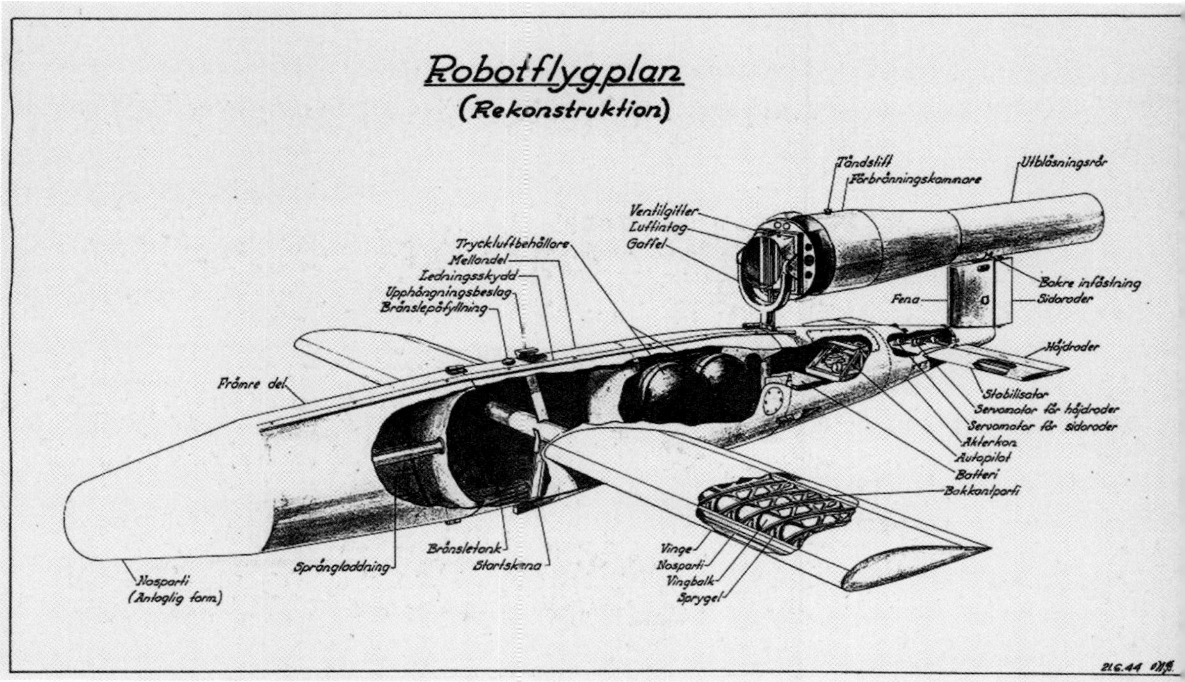

A German sketch of a V-1 flying bomb.

Three photographs showing plans for the construction of the V-1 flying bomb.

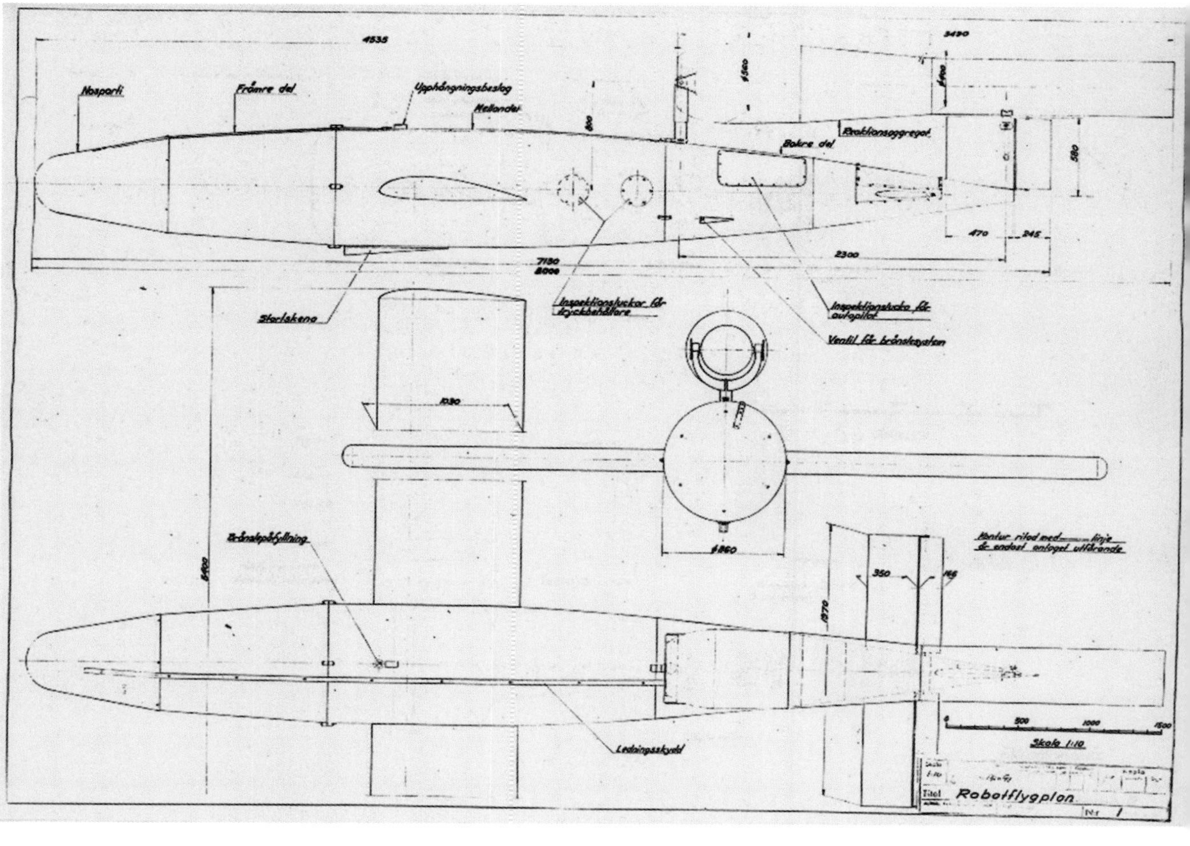

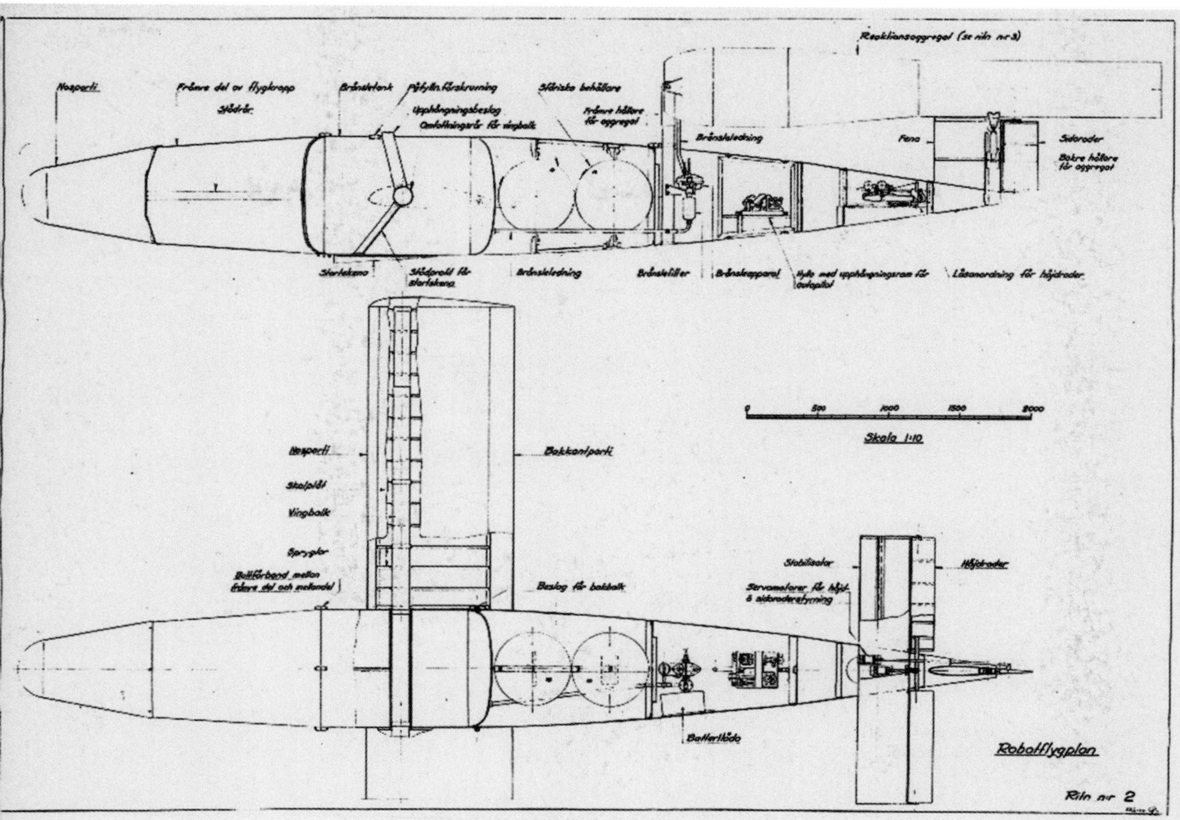
Reaktionsaggregat (se ritn nr 3)
Nosparti
Främre del av flygkropp
Stödrör
Bränsletank
Påfylln. förskruvning
Upphängningsbeslag
Genomföringsrör för vingbalk
Sfäriska behållare
Främre hållare för aggregat
Bränsleledning
Fena
Sidoroder
Bakre hållare för aggregat
Startskena
Stödprofil för startskena
Bränsleledning
Bränslefilter
Bränsleapparat
Hylla med upphängningsram för autopilot
Låsanordning för höjdroder
Skala 1:10
Nosparti
Bakkantparti
Skalplåt
Vingbalk
Spryglar
Bultförband mellan främre del och mellandel
Beslag för bakbalk
Stabilisator
Höjdroder
Servomotorer för höjd- o sidoroderstyrning
Batterilåda
Robotflygplan
Ritn nr 2

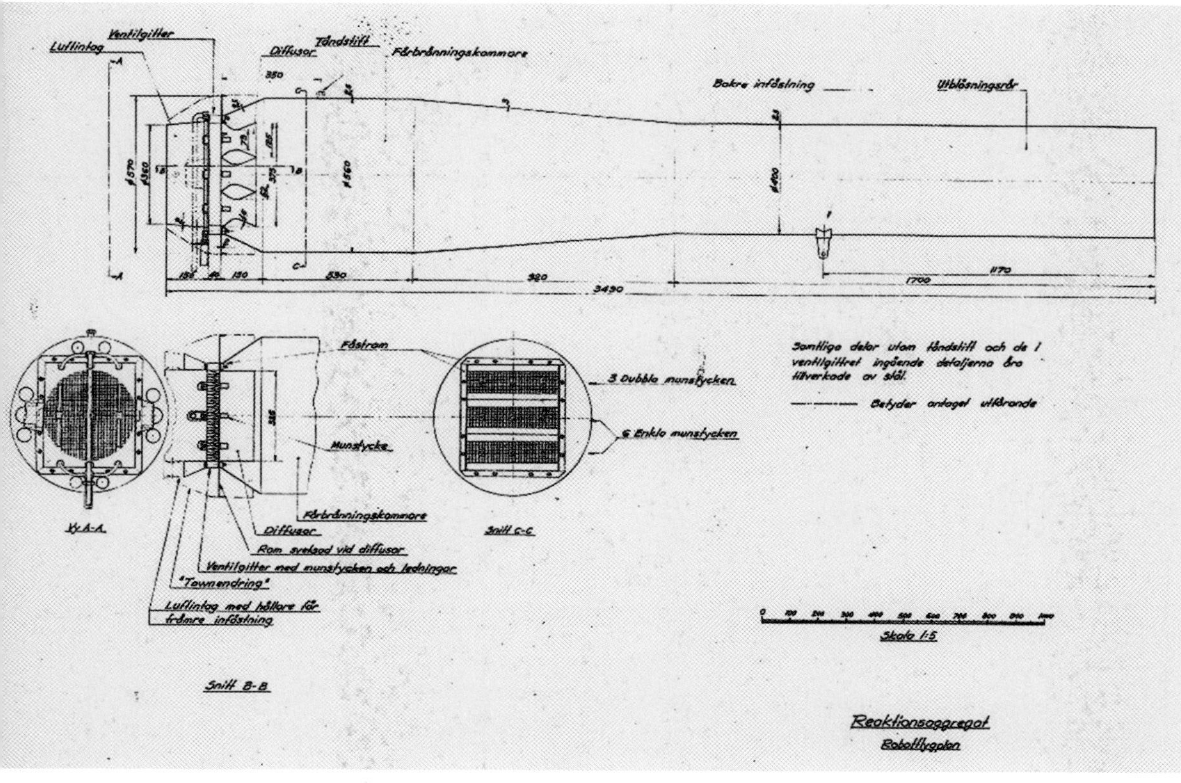
Ventilgitter
Luftintag
Tändstift
Diffusor
Förbränningskammare
Bakre infästning
Utblåsningsrör
3490
1700
1170
Fästram
3 Dubbla munstycken
6 Enkla munstycken
Samtliga delar utom tändstift och de i ventilgittret ingående detaljerna äro tillverkade av stål.
Betyder antaget utförande
Munstycke
Förbränningskammare
Diffusor
Ventilgitter med munstycken och ledningar
"Townendring"
Luftintag med hållare för främre infästning
Vy A-A
Snitt C-C
Snitt B-B
Skala 1:5
Reaktionsaggregat
Robotflygplan

Chapter Two

Slave Labour Production Sites

With plans to develop and mass produce the V weapons, in August 1943 Hitler ordered SS Reichsführer Heinrich Himmler to use concentration camp workers for V-1 and V-2 production. Because Allied bombing had seriously damaged the Peenemünde facility, Nazi leadership decided that plans should be moved to mass produce the 'wonder' weapons in less endangered areas. One site selected was the mountain known as Kohnstein, near Nordhausen in Thuringia. To oversee the creation of this mass production complex, Himmler, Albert Speer, and Reichminister of Munitions Karl Saur agreed on the foundation of Mittelwere GmbH. Its board comprised of Hans Kammler and two of Speers armaments managers, Karl Maria Hettlage and Gerhard Degenkolb. To run the complex, Albin Sawatzki was appointed. Security was given to SS-Obersturmbannführer Helmut Bischoff, who was one of Kammler's staff.

The initial Mittelwerk contract was to produce some 12,000 rockets, which was worth in excess of 750 million Reichsmarks. To achieve this target, the SS had to work quickly, and by the end of August 1943 the first 107 concentration camp prisoners from Buchenwald arrived. A sub-camp was erected known as Arbeitslager Dora. A couple of days later, another 1,223 Buchenwald inmates arrived and this was followed by workers from Peenemünde in mid-October. Over the weeks and months that followed, more prisoners were transported to the area and by Christmas, more than 10,000 workers. Initially, because there was no accommodation for the inmates, the prisoners were forced to be housed inside the tunnels. Conditions inside these tunnels were terrible with no sanitary facilities. The majority of the slave labourers were from Russia, Poland, and France, with hundreds dying from thirst, hunger, cold and being overworked.

Between October 1943 to March 1944, almost 2,900 died at Dora out of the total of 17,500 inmates. In fact, conditions were so bad that the Dora work squads had the highest mortality rate throughout the entire concentration camp system. Prisoners were subjected to extreme cruelty as well, with many of them not seeing natural light for many months. Even Speer, who toured the facility on 10 December 1943 with his staff, was shocked by the dreadful conditions. He later remarked that the tunnels were littered with corpses. The dead had either perished from being worked to death or died of malnutrition and illness. Some who were too weak to perform their duties were either shot or transported to

Majdanek and Bergen-Belsen concentration camps. Yet, in spite of Speer's remarks of horror, a week after his visit he wrote to Kammler, complimenting him on his achievement 'in transforming the underground installation ... from its raw condition two months ago into a factory, which has no equal in Europe and which is unsurpassed even when measured against American standards. I take this opportunity to express my appreciation for this really unique achievement and to ask you also in future to support Herr Degenkolb in this wonderful way.'

By May 1944, Jews were also brought to Mittlebau to work and this included prisoners from Auschwitz-Birkenau as well. In total, some 60,000 prisoners passed through the gates of the complex during its production of the V weapons. In order to accelerate the assembly lines prisoners were forced to dig and expand the tunnels. As more prisoners were drafted to the area, there were other sub-camps established, including Lager Ellrich, Lager Harzungen and Lager Rottleberode.

By May 1944, there was a monthly output of 400 rockets produced, but this was still far below the 1,000 units per month target. On 8 September 1944, the first V-2 rocket that was produced at Mittelbau was successfully launched against London. That month, production had increased to 600 units, and this assembly rate was maintained until February 1945. However, despite this increase in production, materials and fuel were becoming harder to obtain, and coupled with Allied attacks on Nordhausen and advancing US troops, the SS decided to evacuate most of the Mittelbau camps. The evacuation was brutal and thousands died, including many who were forced to board railway cattle cars. In fact, in great haste, the SS left a number of trains containing many thousands of prisoners to die. Others were forced on death marches through the Harz hills to other concentration camps. There were very few prisoners left to be liberated by the US troops when they reached Nordhausen on 11 April 1945.

Some 10,000 deaths were probably attributed to the V-2 programme at the Mittelwerk. Whilst this figure was appalling, the weapon itself caused over 5,000 deaths, primarily in London and Antwerp. Twice as many died producing the missile than being hit by it. Although the rocket was a marvel in modern technology, despite being often flawed and ineffective to change the course of the war, the Nazis had inflicted widespread suffering not just to produce the weapon but to construct the launching sites, too.

(**Opposite, above**) Dora-Mittelbau and prisoners can be seen digging a tunnel under supervision of a German engineer. The production site was established in late summer 1943 as a subcamp of Buchenwald concentration camp, supplying slave labour for extending the nearby tunnels in the Kohnstein and for manufacturing the V-2 rocket and the V-1 flying bomb. By the summer of 1944, Mittelbau became an independent concentration camp with numerous subcamps of its own.

(**Opposite, below**) Dora Mittelbau showing one of the work halls in the underground missile factory. During the initial months, from October 1943 to March 1944, out of a total of 17,500 slave labourers forced to work at Dora, almost 2,900 died.

Dora Mittelbau, showing a heavily camouflaged entrance to the underground missile factory.

Four photographs showing prisoners working on the V-2 missiles' body apparatuses with their fuel tanks on the production line in the tunnels. The V-2 rocket fuel tanks held two main propellants: a mixture of 75 per cent alcohol (either ethanol or an ethanol/methanol blend) and 25 per cent water, and liquid oxygen. These liquids were stored in two large, insulated aluminium-magnesium alloy tanks within the rocket's central section. The alcohol-water mixture served as the fuel, while the liquid oxygen was the oxidizer, both being pumped into the combustion chamber to generate thrust.

NB

Two photographs showing the work halls in the underground missile factory.

Prisoners can be seen moving one of the V-2s along a production line in a tunnel.

Prisoners wiring one the V-2 missiles in July 1944.

Two photographs showing prisoners producing casings for the V-2 missiles in hall no. 31 in July 1944. The V-2 missile's main body or casing was primarily made of thin sheet steel, welded and braced over a wooden framework in some areas, while the warhead casing was made of thick, mild steel. The central section featured aluminium-magnesium alloy propellant tanks, and the tail held the steel motor components and pumps with an aluminium bronze alloy coating.

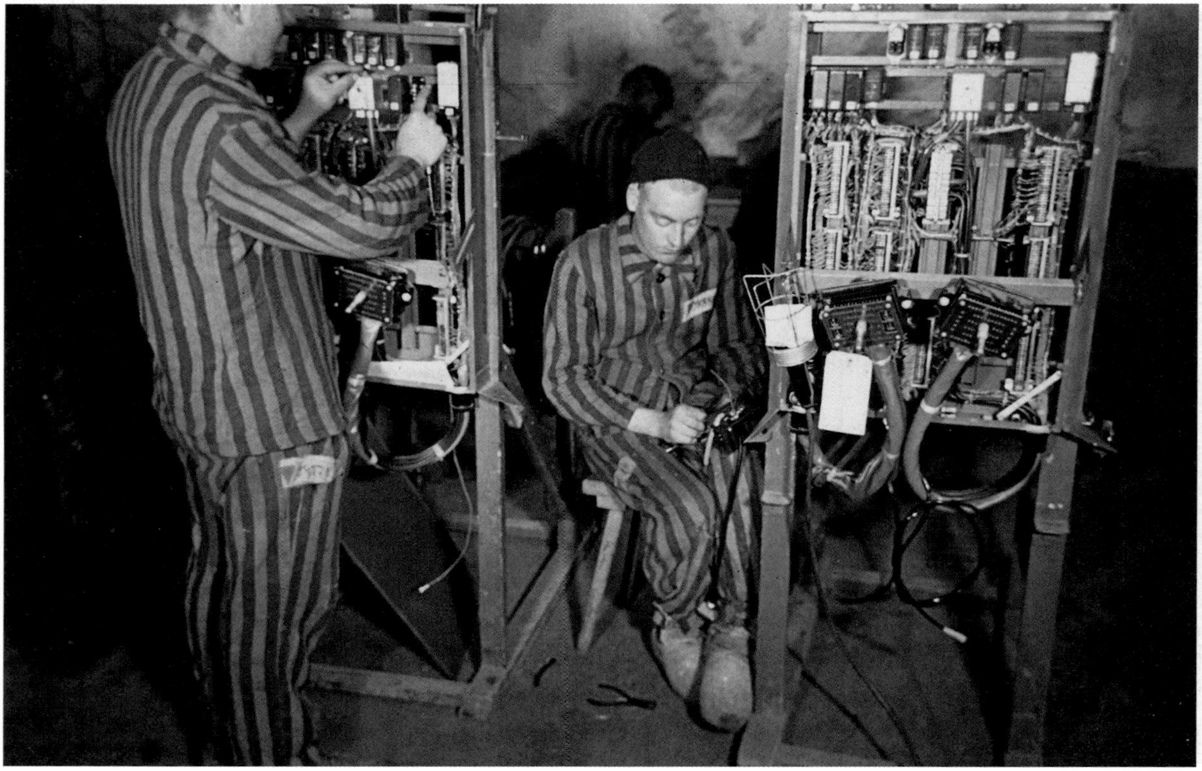

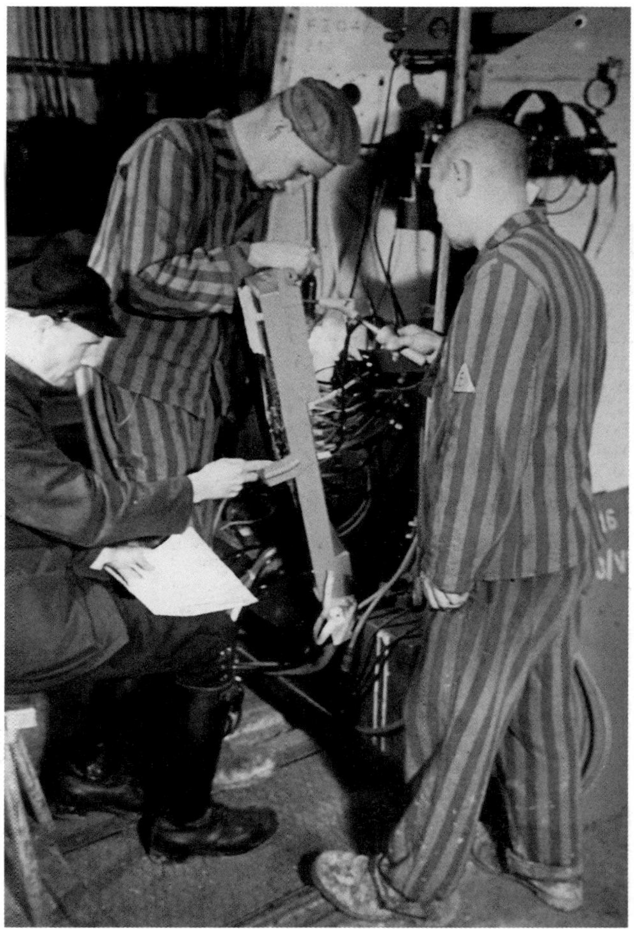

Two photographs showing prisoners (the man on the right is French in the first photograph and in the second image, the man on the right is probably Jewish). They are putting together circuit panels for V-2 missiles in July 1944. The V-2 missile did not use circuit boards or 'circuit panels' in the modern sense; instead, its electrical and guidance systems relied on vacuum tubes, relays, gyroscopes, and primitive electronic components like resistors, capacitors, and inductors to control its flight path and launch sequence. These systems were housed in the rocket's instrument section and were a key part of its 'fully electronic active control system' that became the foundation for modern analogue flight control systems.

(**Opposite, below**) Prisoners can be seen riveting together V-2 casing. The casing was mainly constructed from welded and riveted, thin sheet steel, with some sections reinforced with a wooden framework. It housed the rocket's warhead, instrument section, fuel and liquid oxygen (lox) tanks, and the motor at the tail, which propelled it into the air.

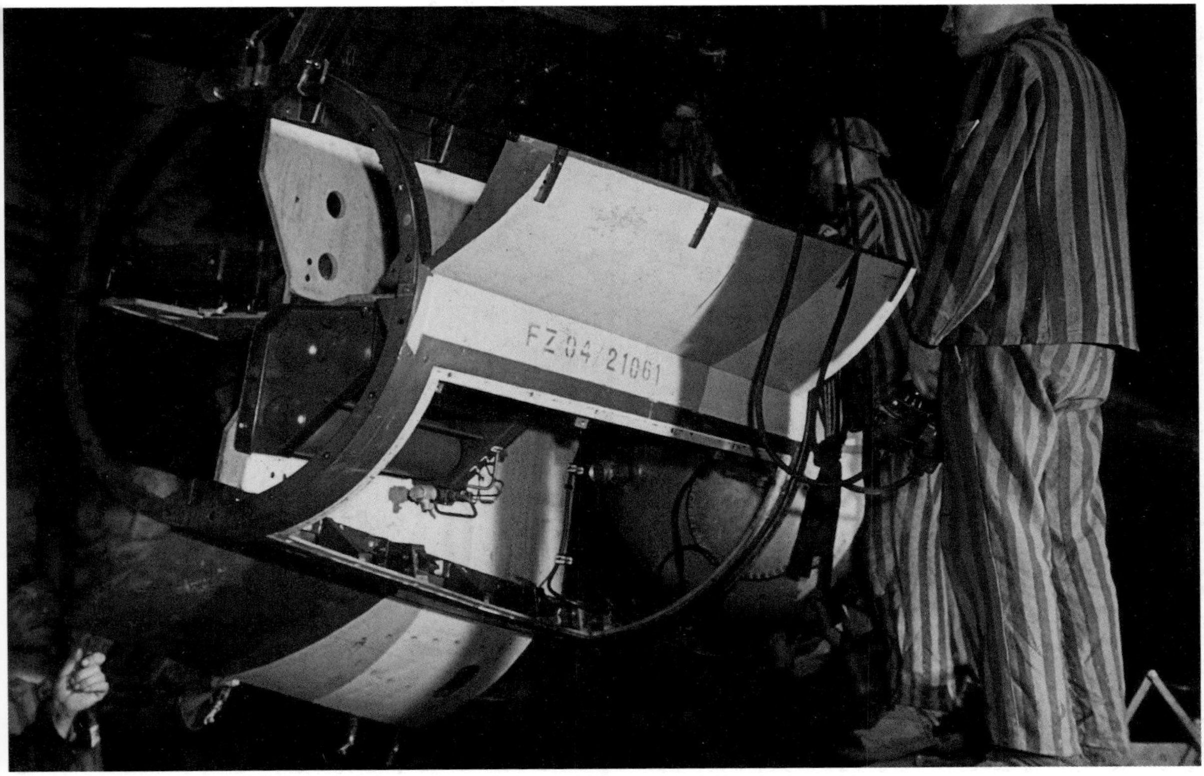

(**Above**) Prisoners, probably Jewish, are putting together a missile head with a flight control system of a V-2 missiles, July 1944. The V-2 was equipped with the world's first fully electronic active control system and a fully inertial navigational system. It utilized a triple-axis stabilized platform that incorporated two longitudinal accelerometers and one lateral accelerometer to provide stability and direction during flight. The control system comprised of what was termed the 'Mischgerät', which was an early electronic computing device that implemented the V-2's stability control equations.

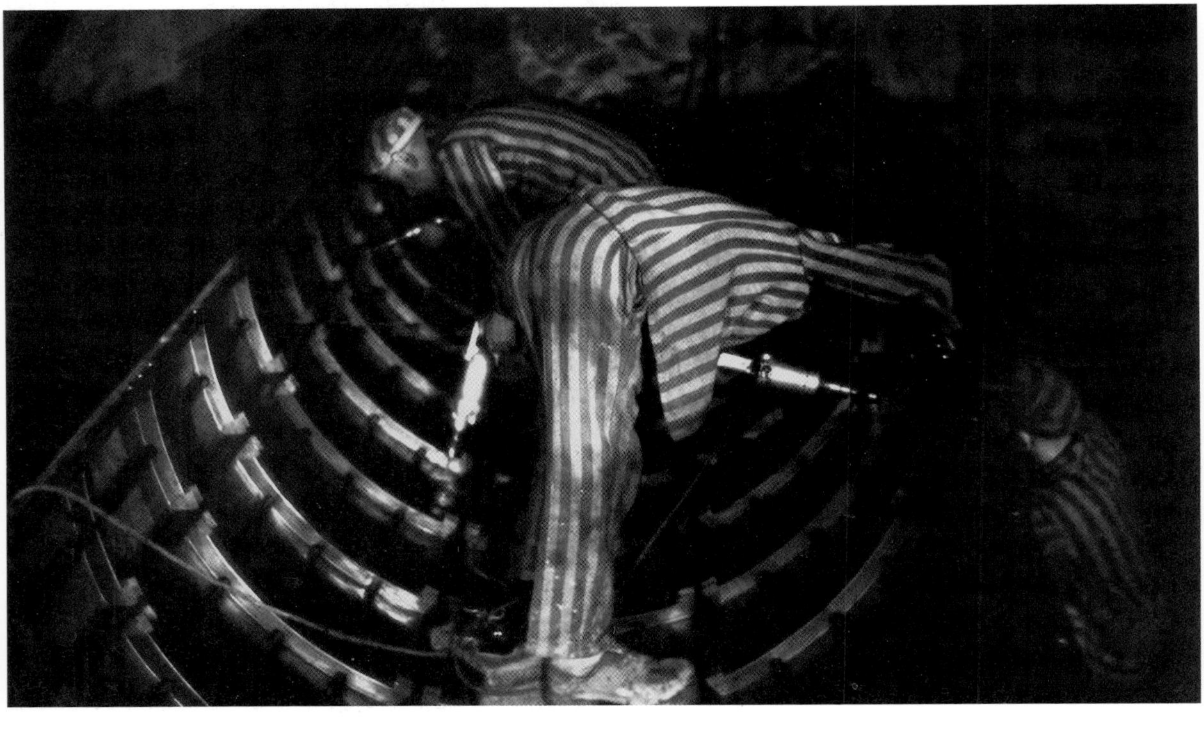

A prisoner and German technicians are attaching the V-2 missile's flight control systems to the missile's head, July 1944. The nose of the V-2 had a contact fuse for the 1-ton warhead, and behind it was the compartment for the analogue guidance computer and gyroscopes that controlled the missile's flight via graphite vanes in the thrust stream.

Taken in July 1944 showing the photographer Walter Frentz with his camera and the special lighting equipment needed for filming inside the factory's tunnels.

The huge hall 41 where the final inspections of V-2 missiles were undertaken before they were transferred to the Wehrmacht.

Two photographs showing prisoners moving a completed V-2 missile, which is being moved with a crane to the inspection hall 41 in July 1944. The assembly process of the V-2 required meticulous work, but the rocket was not built with fully interchangeable parts. Key components, such as combustion chambers, fuel pumps, and valves had to be carefully matched and calibrated for each rocket. This meant that every engine had to be test-fired and certified before final assembly. The Mittelwerk factory comprised of 2.5 miles of tunnels, which included a main assembly line and forty-six connecting side tunnels. The rocket inspection areas were located within these side tunnels, such as Gallery 41, which contained vertical scaffolding for checking the rockets.

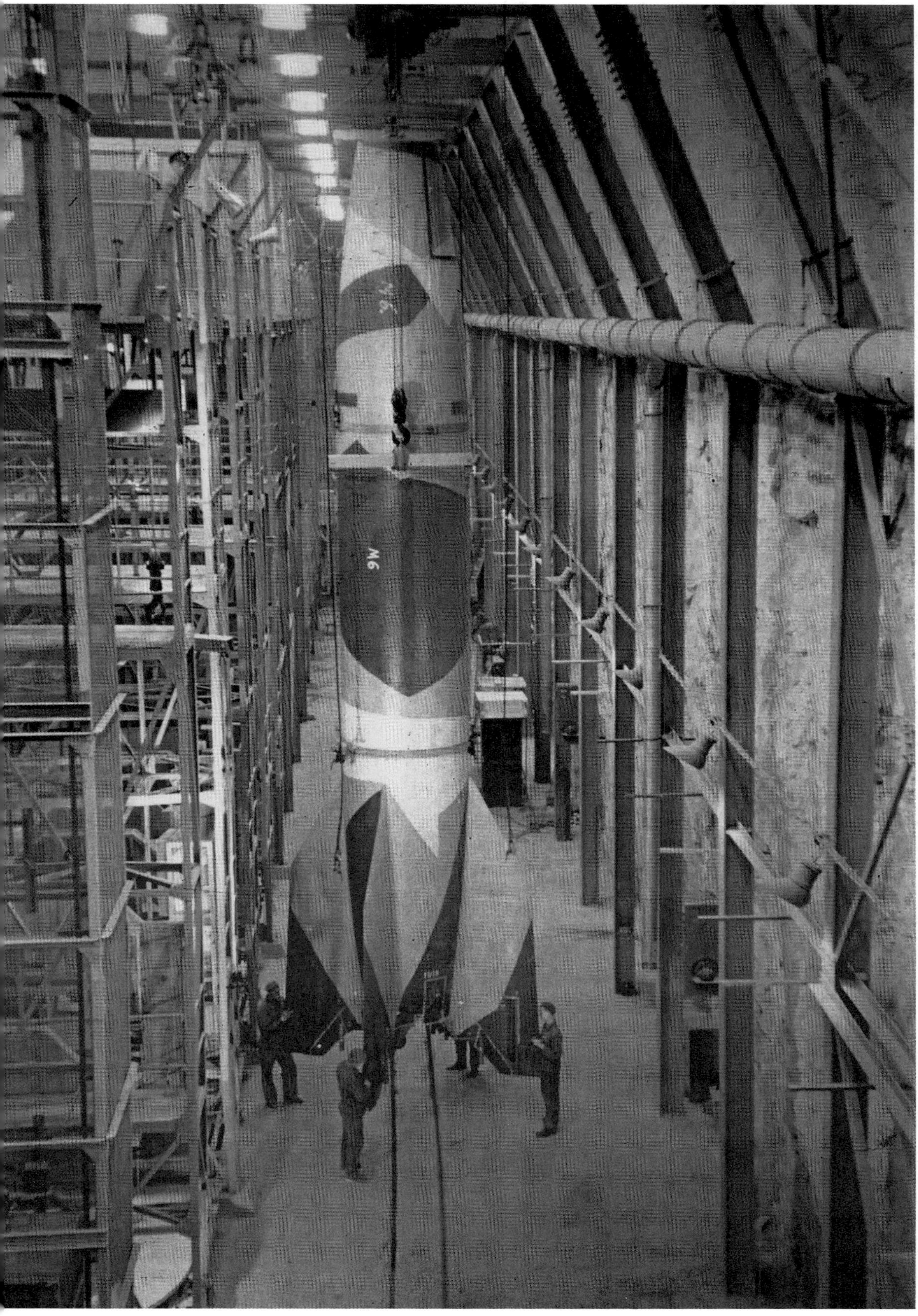

(**Above**) A German technician with inspection equipment calibrating parts of the V-2 missiles in one of the production halls.

(**Opposite, above**) V-2 engine assembly line. The V-2 rocket engine was the world's first large-scale, liquid-fuelled rocket engine. A revolutionary achievement, it propelled the V-2 ballistic missile and laid the foundation for modern rocket propulsion, contributing directly to both Cold War ballistic missile development and the American space programme. The engine's turbo pump produced over 500 hp and rotated at up to 5,000 rpm and the core of the engine was designed to withstand extremely high temperatures. To prevent the high temperatures from melting the engine, two innovative cooling methods were used, called regenerative cooling and film cooling.

(**Below**) A prisoner and a German technician attaching a tail apparatus to the engine system and the body of the V-2 missile in July 1944.

Prisoners working on the tail fin section of the V-2. The V-2 missile had four large fins for stabilization, each featuring both external and internal control vanes that were managed by the rocket's analogue computer to maintain direction and control its trajectory. These control vanes, or 'wind vanes' and 'turbines', were powered by the rocket's jet exhaust to provide steering in pitch, yaw, and roll, a key feature in its early, fully electronic active control system.

A V-2 missile leaving the assembly line. V-2 rockets departing the assembly line involved a complex, large-scale process. All completed V-2s were moved along Tunnel B's railway tracks as a main part of the assembly process, exiting the factory on their way to being launched against Allied targets.

Unfinished V-1 missiles in the underground factory. The V-1 flying bomb production involved mass assembly lines at facilities like the underground Mittelwerk complex and the Volkswagen plant at Fallersleben, using slave labour. Germany produced over 30,000 V-1s, with assembly taking around 350 hours each. This was a massive undertaking, using both large underground sites and smaller, dispersed units to evade Allied bombing, ultimately leading to the deaths of many forced labourers.

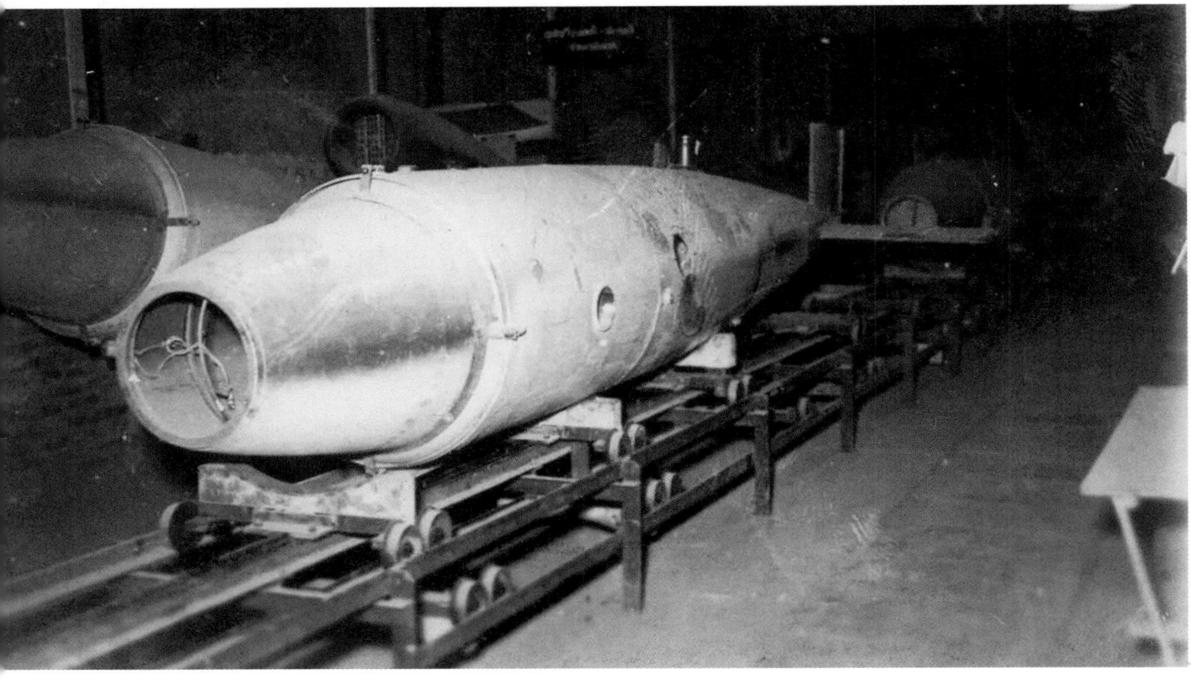

Niedersachswerfen production facility showing V-1 flying bomb casings. The flying bomb's casing was constructed primarily of welded sheet steel, while its wings were made of a simple plywood-and-steel design. The use of these readily available and inexpensive materials was a strategic choice by the Germans, whose industrial base was suffering from resource shortages late in the war. The V-1 was a very cost-effective weapon. A key advantage was its relative simplicity and lower material costs compared to traditional aircraft, allowing for a more cost-effective response to Allied air power.

V-1 flying bombs on an assembly line. The Mittelwerk factory V-1 assembly took place primarily in the vast Mittelwerk underground plant. Whilst building these weapons, workers faced a brutal and inhumane existence, suffering from starvation, torture, frequent executions, and exposure to constant noise, dust, and noxious gases in cramped underground tunnels.

April 1945 and a distressed, but relieved survivor can be seen next to a pot of potatoes at Nordhausen. American troops, specifically the 3rd Armoured and 104th Infantry Divisions, entered the area near Nordhausen where they were unprepared for the scene they found.

Mittelbau Dora and the corpse of a female victim can be seen after the liberation. American troops first discovered the horrors of the Dora concentration camp, where they found 1,300 to 2,500 corpses along with a few thousand starving survivors. The camp's prisoners, mainly from Eastern Europe, were forced to build the V-weapons in brutal conditions.

At Mittelbau Dora and the corpse of a young victim can be seen after the liberation.

Mittelbau Dora and the corpses of camp victims can be seen prior to their burial, after the liberation.

Remains of a captured V-2 fuselage complete with engine can be seen in what appears to be Tunnel B on the railway track.

US soldiers can be seen with an abandoned V-1 flying bomb.

US soldiers can be seen next to an abandoned rail wagon at Nordhausen. In the days leading up to the camp's liberation, the SS evacuated most prisoners on 'death marches' or crammed them into rail wagons. Many of these prisoners died of exhaustion, starvation, or were murdered by the guards.

Dora-Mittelbau crematorium was built in the autumn of 1944 and was in operation around the clock. At least 5,000 of the 20,000 prisoners who died at Mittelbau-Dora were cremated there. Yet, in spite of the brutal forced labour, the capture of Mittelwerk was a pivotal moment that jumpstarted American and Soviet rocket programmes. Wernher von Braun would go on to lead the development of the Saturn V rocket for NASA.

Chapter Three

Secret Installations

Kraftwerk Nord West (North-West Power Plant)

As Watchel hastily began preparing his Flakregiment for attack, all along northern France thousands of labourers were building and erecting mobile launch sites, and massive reinforced concrete buildings to fire both the V-1 and V-2 rockets. This huge building programme had been ongoing since early 1943, and some of the larger installations as far back as autumn 1942. By September 1943, fifty-eight of the first sixty-four sites had nearly been completed, together with a further thirty-two reserve sites spread throughout Normandy, Artois, and Picardy. It seemed that no French wood, no farmstead, no valley was safe from enemy aerial reconnaissance. The allies charted every potential launching site and bombed it with precision. Even one of the most secret installations hidden near the southeast fringe of the forest of Eperlecques, a mile from the town of Watten and 6 miles north of St Omer, received the attentions of enemy reconnaissance.

This huge complex was codenamed Kraftwerk Nord West or 'North-West Power Plant', or Blockhaus d'Éperlecques and also referred to by the locals as the 'Watten Bunker'. It was built as a primary base to launch the V-2 rockets and destroy London. The site was designed in January and February 1943 by engineers from the Peenemünde research facility and the Organisation Todt. On 25 March 1943, blueprints for the construction were submitted to Hitler who authorised its construction immediately.

Some 6,000 labourers from Building Battalion 434 under the supervision of Organization Todt were given the task to construct the site. The workforce comprised of various German specialists who assigned, by force, conscripted Frenchmen from the Service du Travail Obigatoire (STO). In order to supplement the slave labour, Dutch, Belgium, French, Polish, Czech and Soviet PoWs, including civilian conscripts, were made to work. Non-German workers forcibly lived in two camps under Organisation Todt near the village of Éperlecques. This was known as the Watten Zwangsarbeitslager 62 (Forced Labour Camp 62).

The labourers worked under very harsh, strict conditions in 12-hour shifts of 3,000–4,000 men, with three 20-minute breaks during each shift. Construction was seven days a week and throughout the night. Huge floodlights could be seen from the surrounding area as work continued around the clock. In order to feed massive amount of power into the complex, the Germans built a bombproof power station with a 2,000 hp (1.5 MW) generating capacity.

All building materials were brought into the area by barges, trains, and trucks which were unloaded onto a Decauville narrow-gauge railway directly to the construction sites. It was also planned that missiles, warheads, and other components for the V-2 would be shipped to the bunker station and transported on trucks into the main area of the complex. Nearby, an old quarry at Wizernes, codenamed Schotterwerk Nordwest (Gravel Quarry Northwest), was converted into a massive storage facility for the Watten facility.

For weeks, construction of the complex continued on the windowless bunker. The size of the building was on a massive scale. The bunker stretched 300 feet long, 200 feet wide and 50 feet high, extending over 20 feet underground, which was to combine launching with rocket storage facilities and a self-contained liquid oxygen plant so that it enabled crews to fire a continuous stream of rockets across the Channel. The bunker comprised of three main sectors. The main part of the building housed the LOX plant and a vault where missiles would be assembled and prepared. Its walls were up to 23 feet thick. When the rockets were assembled in these special assembly halls, they would then be raised into a vertical position, fuelled and armed. Once they were armed, they would then be slowly moved to either end of the building through pivoting doors 59 feet high. From here, they would exit through the south face of the bunker installation and moved on special tracks to the launch pads. All launches were directed by a special control tower that was located at the centre of the south side of the bunker overlooking the launch pads. Supporting the control tower, a giant Wurzburg radar system was installed to follow the trajectories of V-2s being launched.

Beneath this massive structure of concrete and steel, a labyrinth of tunnels connected various special stores and blockhouses, and to supply these stores, trains were to enter the building via a spur from the main railway line. The area beneath ground also housed five Heylandt compressors, each capable of producing about 10 tons of LOX per day. Approximately 150 tons of LOX were to be stored in insulated tanks on-site. The bunker complex was planned to store up to 108 missiles and enough fuel to supply three days' worth of launches. The Germans planned to fire up to thirty-six rockets a day from the bunker.

Nearby, a large supply dump was erected at Watten near to the River Aa. This site was eventually to be used to store material required for the V-weapon sites in the Saint-Omer area.

The whole complex soon began to take shape, and due to its sheer size, it had become the main focal point for an Allied air attack. On 27 August 1943, whilst labourers were putting the finishing touches to the building, 185 flying fortresses of the US Eighth Air Force attacked the complex, causing considerable damage. Immediately after the air raid, labourers set to work clearing the debris, and within months had hastily repaired the installation and erected new and stronger structures. But fearing another heavy Allied air raid, the Germans cautiously changed the proposed use of the site to manufacturing liquid oxygen, to fuel the rockets which were moved elsewhere to be launched.

Within weeks of the US air attack, an alternative site for bombproof launches had been found. Surprisingly, it was situated just south of the Watten V-2 complex in a quarry at Wizernes. It was code named Bauvorhaben 21 or Schotterwerk Nordwest (Northwest Gravel Works).

Six photographs showing the former rocket facility known as Blockhaus d'Éperlecques which is now a museum. This massive bunker installation was constructed in the forest of Éperlecques, near the town of Watten in the Pas-de-Calais region of France. The site was chosen for its proximity to railways, canals, and power lines, while also being far enough inland to be protected from naval artillery. The facility was intended to house over 100 V-2 missiles at a time and launch up to 36 per day. *(Blockhaus d'Éperlecques Museum)*

Showing a mock-up of a V-2 rocket erected vertically in one of the launching halls.
(Blockhaus d'Éperlecques Museum)

One of the gigantic concrete halls inside the rocket facility. *(Blockhaus d'Éperlecques Museum)*

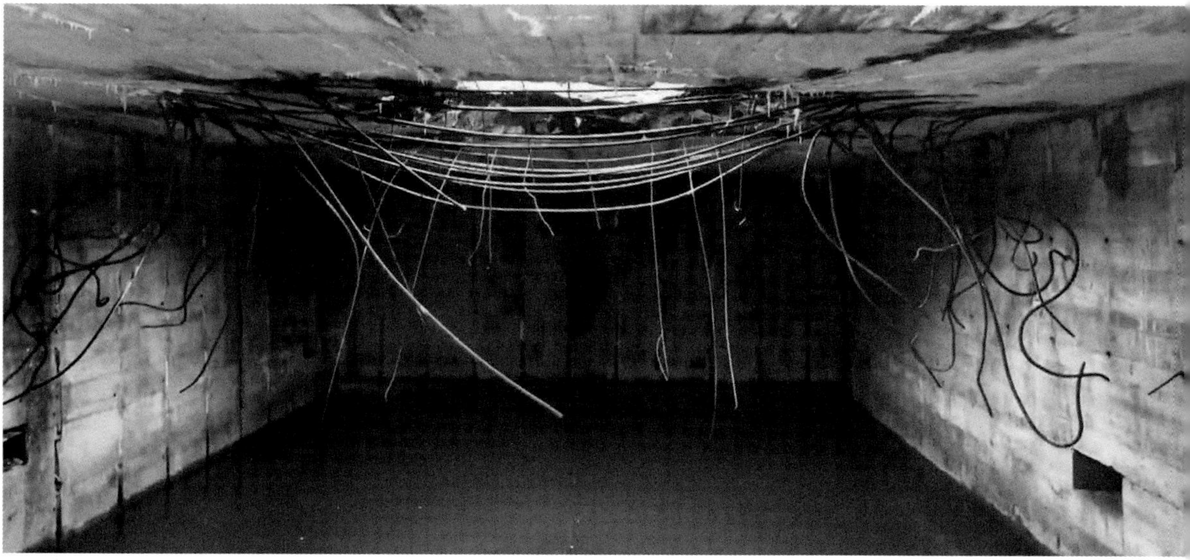

Damage made to part of the bunker installations by a Tallboy earthquake bomb. Intense Allied aerial attacks, which were part of Operation Crossbow, heavily bombed the facility, causing significant damage and preventing its completion. While parts of it were eventually finished to produce liquid oxygen, the bunker was never used for its intended purpose. It was captured by Allied forces in September 1944. *(Blockhaus d'Éperlecques Museum)*

Bauvorhaben 21 (Building Project 21)

On 30 September 1943, Xavier Dorsch, chief engineer of the Organisation Todt met with Albert Speer and Hitler to discuss a proposal for a replacement V-2 rocket launch facility. Dorsch proposed to adapt the nearby quarry site, which was situated southwest of Saint-Omer and some 7-miles south of the Watten bunker, near the river alongside the Boulogne–Saint-Omer railway line, very close to Wizernes station. Inside this quarry, Dorsch showed Speer and Hitler plans for a huge bombproof storage bunker complex, which could be used for assembly and launching V-2s. The bunker would be constructed by a method German engineers called 'Verbunkkerung'. This planned impregnable fortress was to be protected by a 1-million ton dome that would be constructed on a hillside overlooking the quarry. The dome would be protected around its perimeter by a bombproof 'skirt' or Zerschellerplatte of steel-reinforced concrete, 46 feet wide and 6.6 feet thick. This was supported by a series of buttresses close to the dome structure. Near the dome, there was a concrete cubed shaped structure on the top of the hill that was intended as a bombproof outlet for an air conditioning and ventilation shaft. Plans were also drawn up for a concrete structure that would serve as an observation and control tower. There would also be an underground facility built on the western side of the quarry to provide the facility with a hospital and offices for the engineers. Unlike the Éperlecques, Blockhaus designers decided not to include electricity at Wizernes. Instead, power was provided by a connection to the main electric grid.

Plans for the new V-2 facility were impressive and the engineering, exceptionally advanced. Hitler was so overwhelmed by its size and capabilities he immediately ordered its construction, and in November 1943, thousands of labourers began to work around the clock building, digging, and excavating. Over half of the workforce were Germans comprising of skilled workers, such as miners from Westphalia, who were recruited to excavate the tunnels and erect the dome. The remainder were slave labour, mainly Frenchmen conscripted by the Service du travail obligatoire (STO), plus Soviet PoWs. The project was overseen by a number of German construction companies, with the main contractors being Grossdeutsche Schachtbau, Tiefbohr GmbH and Philipp Holzman A.G. of Frankfurt am Main.

Beneath this huge dome, erected by early 1944, tunnels were excavated into the chalky hillside that led to two massive rocket-high doorways hidden in the quarry itself. The rockets would be transported to the Wizernes bunk via a small gauge railway line that entered the quarry and then a tunnel in the hillside. The rail line codenamed Ida, was to be built on a curving path that would connect it with both the east and westbound main line railway, allowing trains to run straight through the facility without needing to reverse or be turned around. This rail line would serve as the main unloading station, where missiles and supplies would be offloaded onto large trolleys that would be moved quickly through the connecting tunnels and galleries nicknamed 'Mathilde' and 'Hugo'. Each of the main tunnels had a number of unnamed side tunnels which were up to 300 feet long. The central part of the facility was a huge octagonal rocket-preparation chamber directly under the dome. This was still under construction, but never completed. It was huge and was 135 feet in diameter and up to 108 feet high. A series of floors were also being built up the side of the chamber.

From here, in the relative safety of this vast underground cavern, V-2s were to be assembled and serviced. They would be then taken through a set of 55 feet high bombproof doorways made of steel and concrete along tracked passages to the launch pad named Gustav on the southern pad or Gretchen on the northerly located pad, and fired from the quarry towards London. Hitler wanted to see London totally decimated by the V-2s and there were even plans to target New York, as the facility was large enough to build a proposed V-2 twice the size known as the 'American Rocket'. Against London, it was intended that at least fifty rockets a day could be fired from the complex. But in March 1944, these plans were interrupted by a heavy Allied bombing raid. Yet despite this setback, construction continued. In fact, due to Allied air attacks, the dome's designer, Todt Organisation engineer Werner Flos, had already devised a plan where the dome would be built first, constructed on flat ground, and then the soil underneath excavated so construction work would not be hindered below and would protect labourers against aerial attacks. A circular trench had been dug on the top of the hill above the quarry to an outside diameter of 276 feet. The dome was then constructed within this trench.

However, in spite of the progress of the construction of the launching site, the Allies continued their bombing campaign of the facility. By the early summer of 1944, Allied air intelligence had become so intense that nothing along northern France was safe from bomber attacks. In June and July 1944, the RAF began bombing raids against the complex using massive 5-ton bunker busting Tallboy bombs. As a result of these attacks, the external construction works were completely decimated by the bombing. Other attacks caused widespread damage beside the dome, blowing out the complete quarry cliff face and burying the entrances to the Gustav and Gretchen tunnels. The entrance to Sophie was also buried, leaving Ida as the only entrance to the complex. The dome was undamaged but the buttresses supporting the protective Zerschellerplatte were blown off and landed in the quarry.

With the Wizernes facility damaged, any hopes of launching V-2 rockets were dashed. The two enormous bunker sites in the Cherbourg peninsula at Martinvast and Sottevast, which had been under construction to launch attacks by V-2 rocket on the towns and ports of southwestern England, were also attacked. Although only half completed, the installations were to have a similar use as the intended Éperlecques Blockhaus site. Rockets were to be serviced, fuelled, and tested in a bombproof construction, again built on the 'Verbunkkerung' method. But like the rest of the V-2 sites, it was bombed and five months later in July 1944, it was captured by the advancing American forces. None of the V-2 bunkers were completed. While they were built to withstand normal bombs, the installations were not designed to withstand Dr Barnes Wallis' special earthquake bombs.

Blueprint cross section of the dome of the Wizernes Bunker Bauvorhaben 21 (*aka* La Coupole).

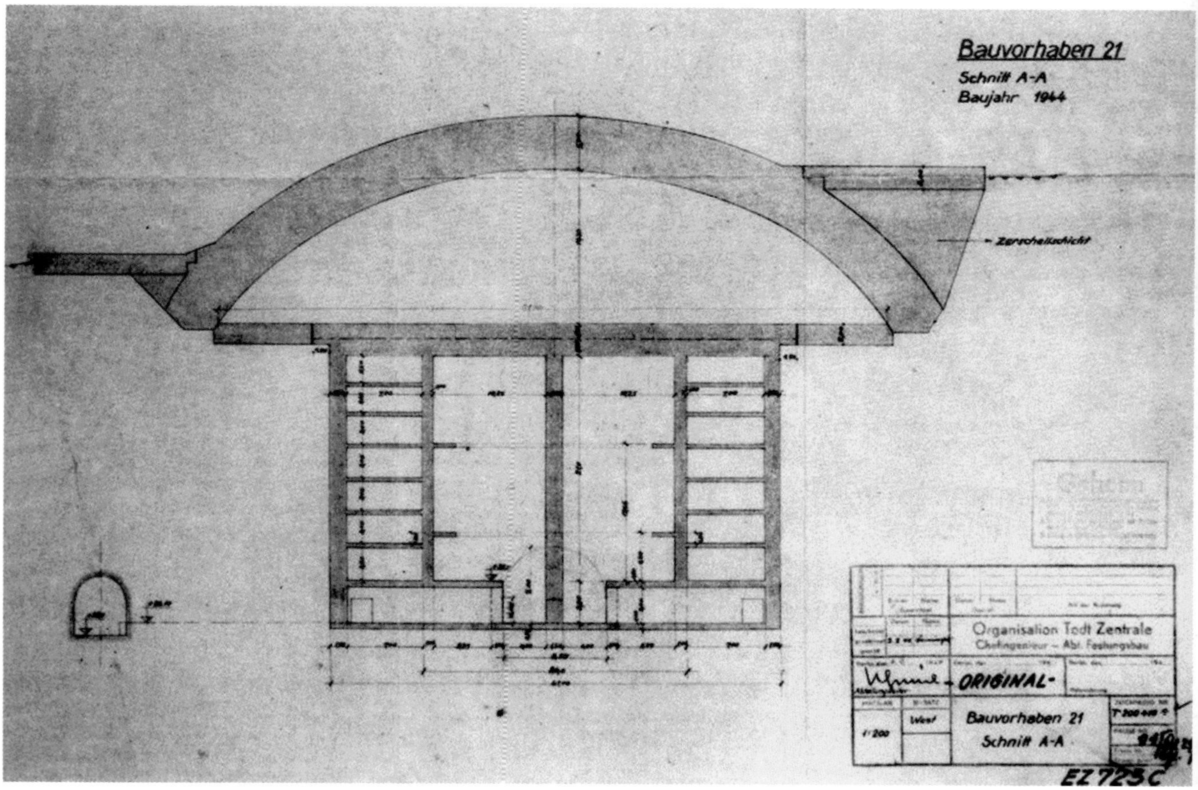

As a result of the extensive amount of damage to the V-2 sites, on 18 July 1944 Hitler abandoned plans for launching V-2 rockets from bunkers, and sent an order authorising the downgrading of the Wizernes site and making it a LOX production facility. However, due to the Allied landings along the shores of Normandy and the advance of ground forces towards the Pas-de-Calais area, the site was completely abandoned.

With the V-2 without bunkers, it was suggested that the rockets would have to be launched, as first suggested, by special mobile launching units.

Aerial reconnaissance photograph of the Wizernes Bunker. This site was built by the Organisation Todt to serve as a central launch and storage site for V-2 rockets targeting London and southern England. The domed structure was situated in a chalk quarry near Saint-Omer in France. The complex was designed with an enormous concrete dome to conceal the underground bunker and its launch preparation chambers.

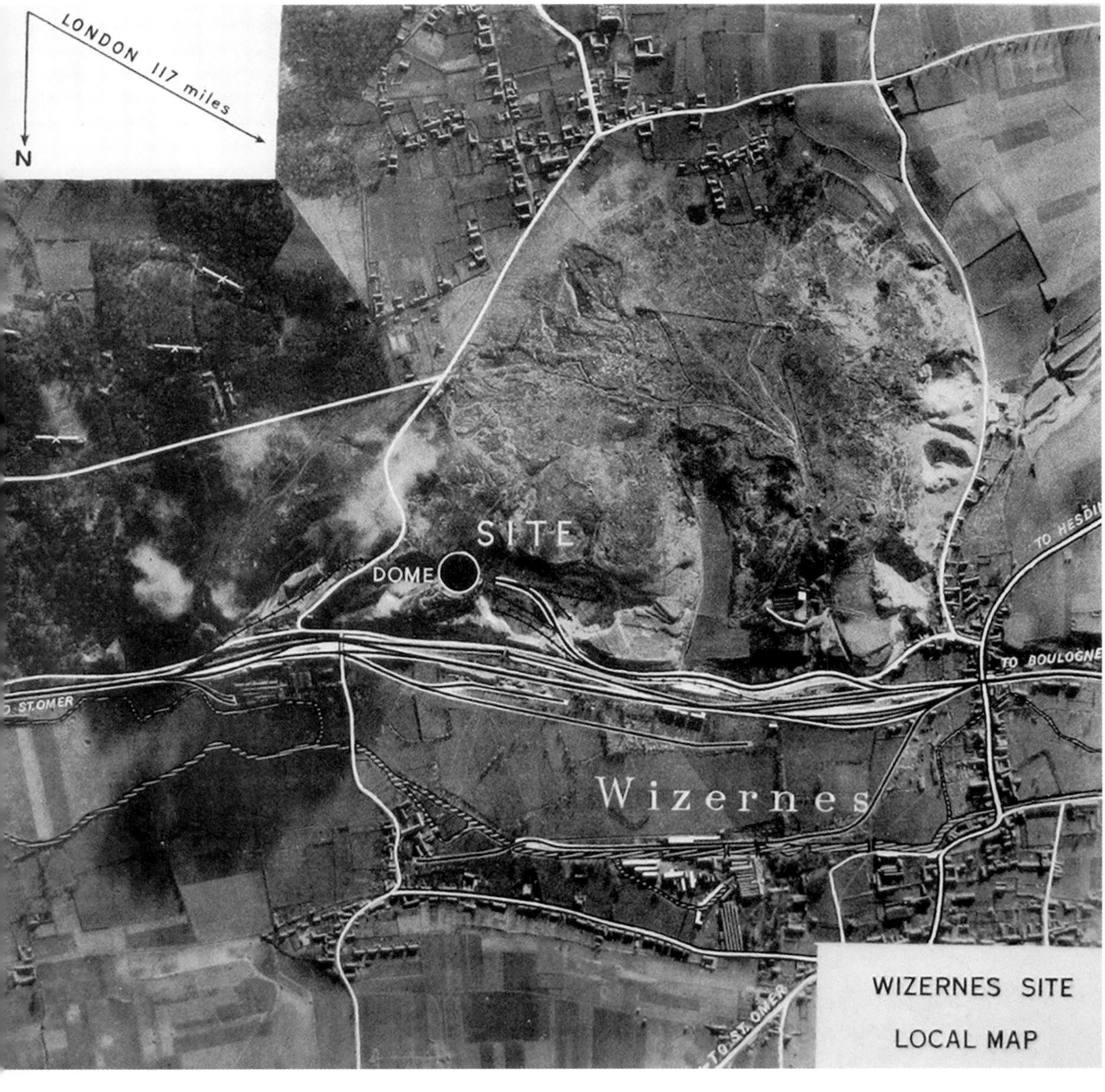

A photograph of the domed concrete structure. The construction of the dome involved a 'tortoise technique', where a thick shell was built and then lifted by hydraulic jacks, allowing for construction within a protected space.

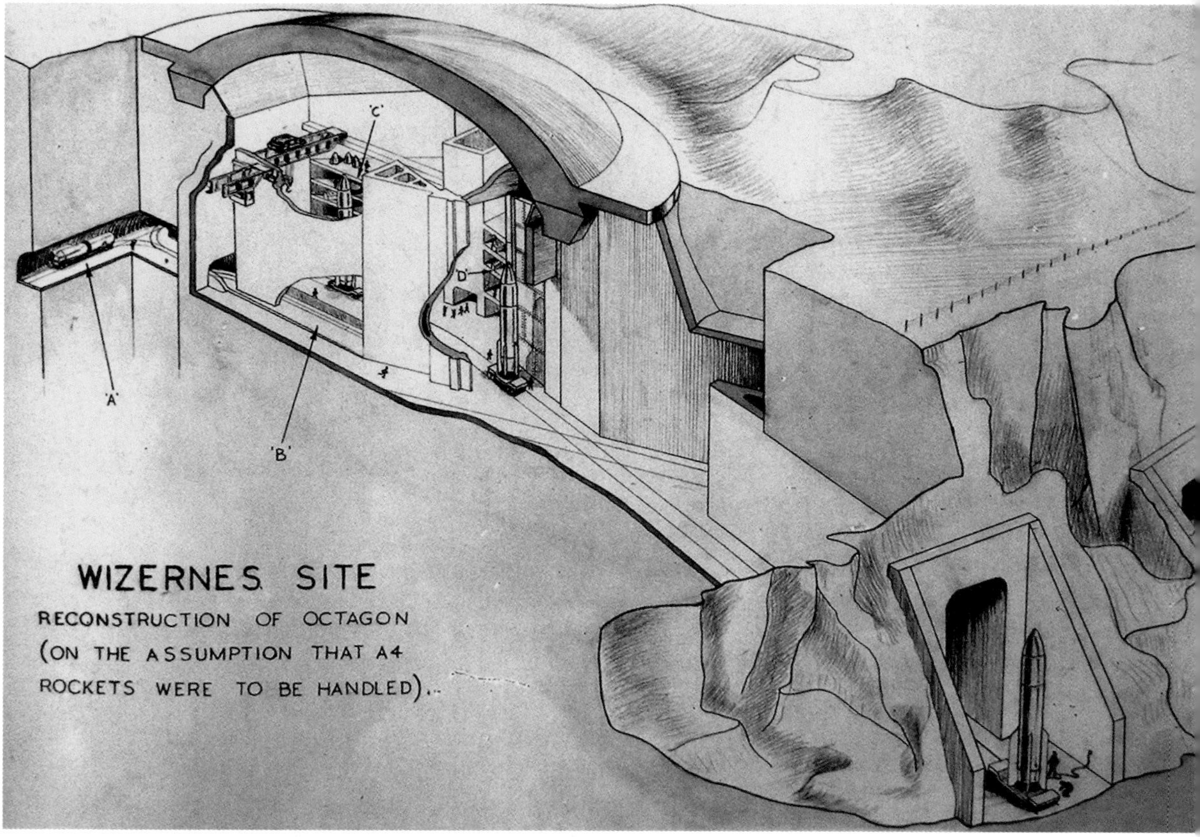

A September 1944 reconstructed drawing of the rocket facility showing how the site would launch the rockets.

Three photographs showing the V-2 bunker at Wizernes and its entrance tunnel including construction inside the dome. The entrance tunnel was part of a planned underground network to transport and launch V-2 rockets. Trains would enter the complex through the tunnel to deliver rockets, which would then be moved into the main preparation chamber beneath the large concrete dome. While the bunker was heavily bombed, the entrance tunnel remained intact, and the site is now a museum and planetarium, offering visitors a look at the underground galleries.

Two photographs showing the entrance of the bunker installation.

Four Allied aerial photographs of the aftermath of bombing of the V-2 bunker at Wizernes. Allied air forces repeatedly bombed the V-2 bunker during the summer of 1944 as part of Operation Crossbow. While the massive concrete dome resisted direct hits, Tallboy earthquake bombs caused extensive damage to the surrounding area and internal structures, leading to tunnel collapses and making the site unusable. The bunker was ultimately abandoned by the Germans in the summer of 1944 and was never used to launch V-2 rockets.

1
2
3
4
5

Three photographs showing the dome facility. Note the series of buttresses closely connected off the dome structure. Also near the dome to the left is a concrete cube-shaped structure on the top of the hill, which was intended as a bomb-proof outlet for an air conditioning and ventilation shaft.

V-1 Rocket Sites

It was not only the V-2 launching sites that had been seriously impeded by the Allied bombing campaign – several V-1 sites had been destroyed before being made operational. The sites most prone for attack were the ingenious V-1 bombproof storage, assembly and launch bunkers, where from beneath the steel and 15-foot carapace, flying bombs were launched into the skies bound for England. Inside the V-1 chamber, which measured about 600 feet long, 48 feet wide and 14 feet high, the flying bombs and ancillary equipment were housed, capable of preparing and firing ninety launches a day at England. These storage buildings were built purely to protect the V-1 launch sites against air attack.

However, despite the Allied air attacks, it did not prevent the V-1 sites launching their 'flying bombs'. Across northern France, in nine general areas, a number of installations, including four launching ramps were aligned towards London and the southern coast of England.

By October 1943, more than 40,000 labourers were working day and night on the construction sites. Weeks later, ninety-five launching sites alone had been erected in various stages of construction. During this period of activity, seventy-two of the ramp sites were located by Allied aerial reconnaissance and fifty-two of them were attacked, with seven completely destroyed. From the air, the launch sites resembled skis lying on their sides, and their characteristics were easily identified from enemy air reconnaissance. The ski shaped buildings, nicknamed by the Allies as 'ski sites', were designed in this particular way to offer additional protection from explosion for the storage of unassembled flying bombs, and also as a shelter for personnel whilst the bomb was fired off. For the crews attacking these installations, German defences were often strong. On approaching a site, airmen became exposed to the full force of its defences. These usually comprised of a number of 3.7cm Flak 43 guns, often sited on the roofs of small buildings, and up to 18 quadruple 2cm Flakvierling 38 guns at any given time. Outside the flak zones were FW 190s patrolling the air, waiting to intercept allied aircraft.

In spite of strong defensive positions, airmen continued to risk their lives. Aircraft lucky enough to escape unscathed from the storm of German flak and fighter aircraft were able to cause considerable damage to the launch sites. Bombing of these sites was violent and remorseless. By January 1944, a quarter of the sites had been put out of action, and more than 20,000 French labourers, nearly half the construction force, had run away. The remaining population at launching zones, which up to now had fired one single 'Retaliation' weapon in anger, had to summon up their reserves of strength in order to continue fulfilling Hitler's objective of destroying London.

In December 1943, General Erich Heinemann, an artillery officer, was given full command of the new LXV Army Corps and its operational formations, Flak-regiment 155 (W) with four Abteilungen, each consisting of four launching and two maintenance and supply batteries – and the senior Artillery Commander 191 leading the V-2 fixed and mobile units.

By order of the OKW, Oberkommando der Wehrmacht or Armed Forces High Command, Heinemann's corps were to prepare the commitment of the V-1 and V-2, and long-range artillery, and direct their employment against England as soon as the weapons were available. But Heinemann was now aware that the 'ski' chain of sites had to be abandoned. The Allies had already dropped thousands of tons of bombs on the now useless sites. It was decided for security reasons to find new but even simpler ways of launching the bombs.

In the intervening months of 1944, whilst Heinemann waited for the first arrival of flying bombs, he spent hours considering the possibility of constructing a new kind of launching site designed and located in such a way which rendered them very difficult to detect photographically. These new sites consisted of a simple metal ramp which could be manufactured off-site and bolted together to previously prepared concrete foundations. No large impregnable buildings were to be constructed; only a number of easily assembled huts would be erected near the ramp. Where possible, the entire area was hidden under natural camouflage afforded by overhanging trees, and extensive use was made of camouflage netting, which sheltered both the huts and subsequent operations. To construct these new sites for the prefabricated catapults and to maintain high levels of secrecy, only German labour was used. Some 10,000 convict workers were drafted in and employed around the clock to work on the sites. Only when the site was completed was the ramp transported to its secret location – which varied from a remote forest to a farm, village, or orchard. Only six days were needed to assemble the sections together and bolt them to the concrete foundations.

As preparation of the sites continued, other necessary equipment on which the Flakregiment crews depended was moved to the sites by railway for immediate installation. Nearby, fuelling areas were built along the railway tracks so that fuel could be pumped from the tank wagons into specially constructed underground storage dumps from which pipes led to fuelling bays bordering the main road. At any one time, four flying bombs could be fuelled whilst they sat on trucks that were to take them directly to the launching sites. In addition to these fuelling dumps, two massive bombproof underground stores were to supply bombs and fuel to all sites south of the Somme. Convoys of lorries were to run each night to large camouflage flying bomb parks and return in the early hours of the morning, while the cargoes they had just delivered were loaded on to specially designed lorries and trailers and delivered the next night to the forward depots. Flying bombs that were scheduled for delivery in the Pas de Calais area would cross the Belgium/German borders by rail and arrive in special marshalling yards near to the border with France. From here, trains were to divide and transport their deadly cargoes to the launching sites in France, and the few that were in Belgium.

Despite the huge undertaking of preparing the flying bomb assault against England, the LXV Army Corps was far from ready. To make matters worse, on 6 June 1944, Heinemann and his staff were given the news that the Allies had begun their invasion of mainland Europe. Confusion and uncertainty swept the German High Command, and through its busy signal traffic emerged an urgent OKW directive informing Heinemann that the bombardment of England was to be brought forward to 12 June. At 5.45pm General Wachtel received code-word 'Rumpelkammer'. Within hours, heavy launching equipment was removed from its specified dumps in Germany and loaded onto trains for the launching zones. The trains, sometimes more than thirty-five wagons long, could only move at night and their loads frequently completed their journey by road, which was often hazardous or incorrectly routed, causing further delays.

Wachtel and his men waited anxiously for the train and lorries to arrive. Not only were they harassed by their superiors, but they also had to contend with the fear of enemy air attacks, and now an even bigger possibility of being overrun by ground forces that were advancing from the shores of Normandy. With these growing concerns, anxious crews were now expected to fire live bombs under operational conditions from untested launching ramps.

Two photographs showing the transportation procedure of moving the V-1 flying bomb to its ramp for launching. The V-1 was moved using special trolleys and handling carts at launch sites and during the final assembly stage. The weapons, also known as 'buzz bombs' or 'doodlebugs', were transported carefully to protect their sensitive guidance components before being hoisted onto a launch ramp. German military personnel used hand-pushed carts and small, wheeled trolleys to move the V-1s around the site. The fuselage and wings were brought to the launch facilities by train or other transport. The wings were typically attached onsite to the main tubular spar.

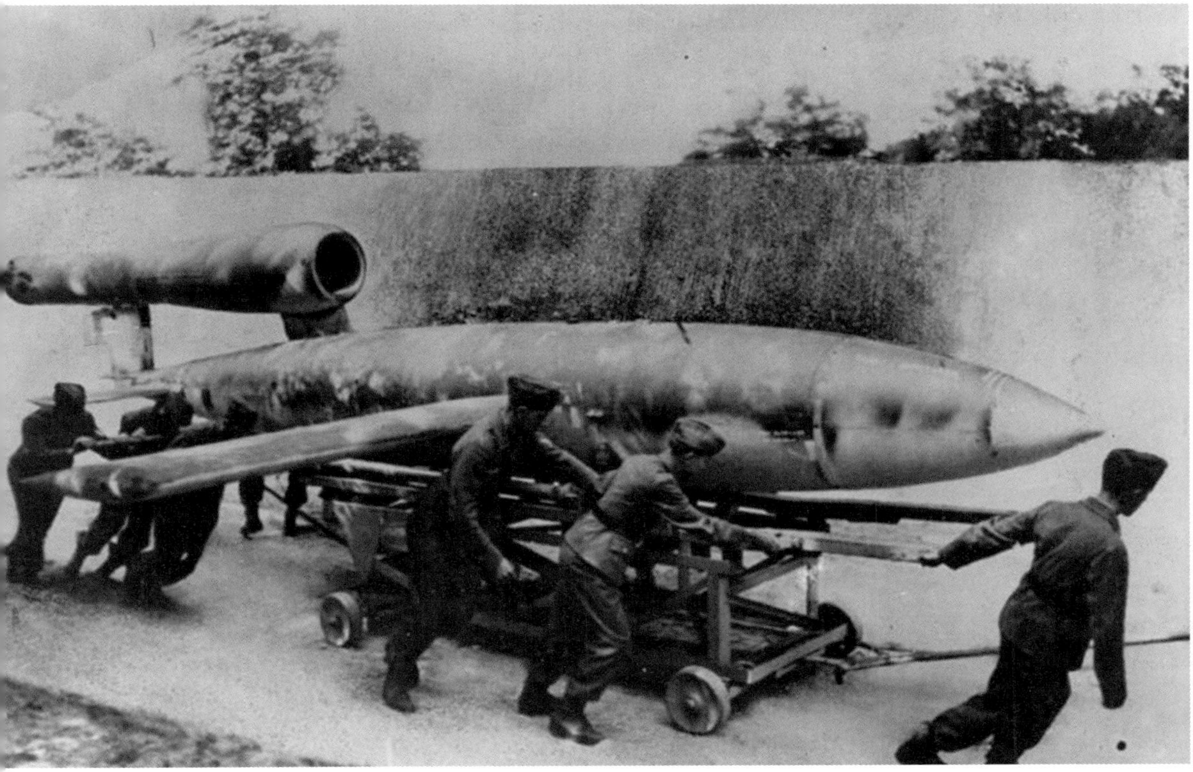

(**Opposite, above**) Former Grandcourt V-1 launch site in France located near Guerville and Grandcourt. Remnants of the original bunkers and bomb craters can still be seen today.

(**Opposite, below**) A V-1 being prepared for launch at Val Ygot near Ardouval, France, about 15 miles south of Dieppe, in the Eawy Forest. Built in late 1943 to launch V-1 rockets at London, the base included bunkers, assembly areas, and a launch ramp. However, Allied intelligence led to repeated bombings of the site, and a V-1 was never successfully launched from Val Ygot before the Germans abandoned it in 1944.

(**Above**) Hidden under camouflage netting is the construction of a large V-1 site. Note the cement mixer and light railway. The layout of these sites followed a standard design for assembly-line preparation of missiles. The layout typically included large, ski-shaped storage bunkers to hold the V-1 missiles, Assembly and workshop buildings, storage bunkers for the hydrogen peroxide and potassium permanganate chemicals that powered the launch catapult, a control bunker for the launch crew, a non-magnetic building to set the missile's gyroscope compass, which was crucial for its directional accuracy. However, the distinctive shape of the 'ski sites' made them easy targets for Allied bombers under Operation Crossbow, which was initiated in December 1943. Many of these sites were severely damaged or destroyed before they could be used.

(**Opposite, above**) The V-1 launch site No.17 near Du Mesnil-au-Val. This launch site for V-1 flying bombs was located near the village of Le Mesnil-au-Val, to the southeast of Cherbourg in Normandy, France. The Germans constructed numerous simple V-1 launch bases throughout northern France. These smaller, basic designs were intended to avoid detection and bombing by the Allies. The sites consisted of a concrete launch ramp and support buildings. At the Mesnil-au-Val site, the remaining structures include support buildings and the road built by the Germans to transport the rockets.

(**Opposite, below**) A V-1 launch site at Vacquerie le Boucq. The site at Vacquerie-le-Boucq included the curved storage bunkers and other concrete structures, such as a preliminary assembly building, fire-control bunker and 'ski' storage structures. The layout of the site is similar to other V-1 facilities in northern France. However, like most of the sites, these were detected and heavily bombed by the Allies before they could become operational. This forced the Germans to switch to smaller, more mobile, and camouflaged launch sites.

(**Above**) The lightly damaged launching ramp of the modified V-1 launching site at Belloy-sur-Somme, near Amiens, France. The site was part of the German Flakregiment 155 and was found by Allied forces after the Germans had retreated. The launch site was mostly destroyed by the retreating Germans, but a prefabricated 'ski-slope' launching ramp remained partly intact. A makeshift wooden derrick was also left at the site.

(**Above**) Former V1 launch site at Val Ygot near the village of Ardouval, in the Normandy region of France. The Val Ygot site was never operational as the launching site was detected by Allied reconnaissance and destroyed by bombing in 1943.

(**Opposite, above**) A photograph taken at the Nevington War Museum showing a reconstructed launch ramp and replica V1 rocket mounted on the ramp in preparation for launching.

(**Opposite, below**) A German diagram showing the Nevington launch site. *(Nevington War Museum)*

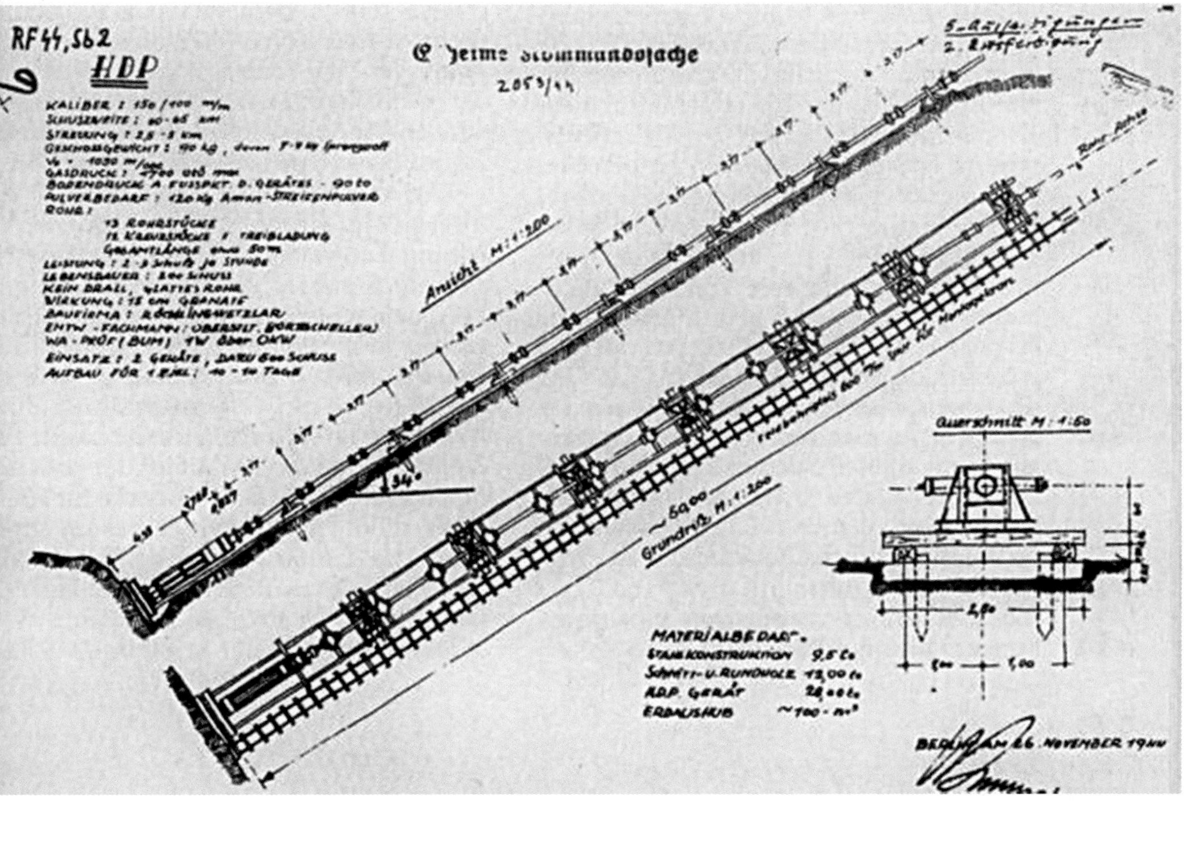

RF44, Sb2
HDP
Geheime Kommandosache
KALIBER : 150 / 100 m/m
STREUUNG : 2,5 - 3 km
ROHR :
13 ROHRSTÜCKE
GESAMTLÄNGE
LEBENSDAUER : 200 SCHUSS
KEIN DRALL, GLATTES ROHR
WIRKUNG : 15 cm GRANATE
WA - PRÜF (BUM) TW über OKW
Ansicht M : 1 : 200
Grundriß M : 1 : 200
Querschnitt M : 1 : 50
34°
~ 60,00
2,80
MATERIALBEDARF :
STAHLKONSTRUKTION 9,5 to
HDP GERÄT
ERDAUSHUB ~ 100 - m³
BERLIN AM 26. NOVEMBER 1944

The most common method of launching the V1s was from a special ramp. The V1 was placed on a steam-powered catapult that rapidly propelled it forward along the ramp, giving it the necessary speed to become airborne. This photograph shows part of a section of the ramp along with a gantry crane used to hoist the flying bomb. *(Museum Blockhaus at Éperlecques)*

A typical V1 site in 1944. *(Cross Bow Committee, April 1944)*

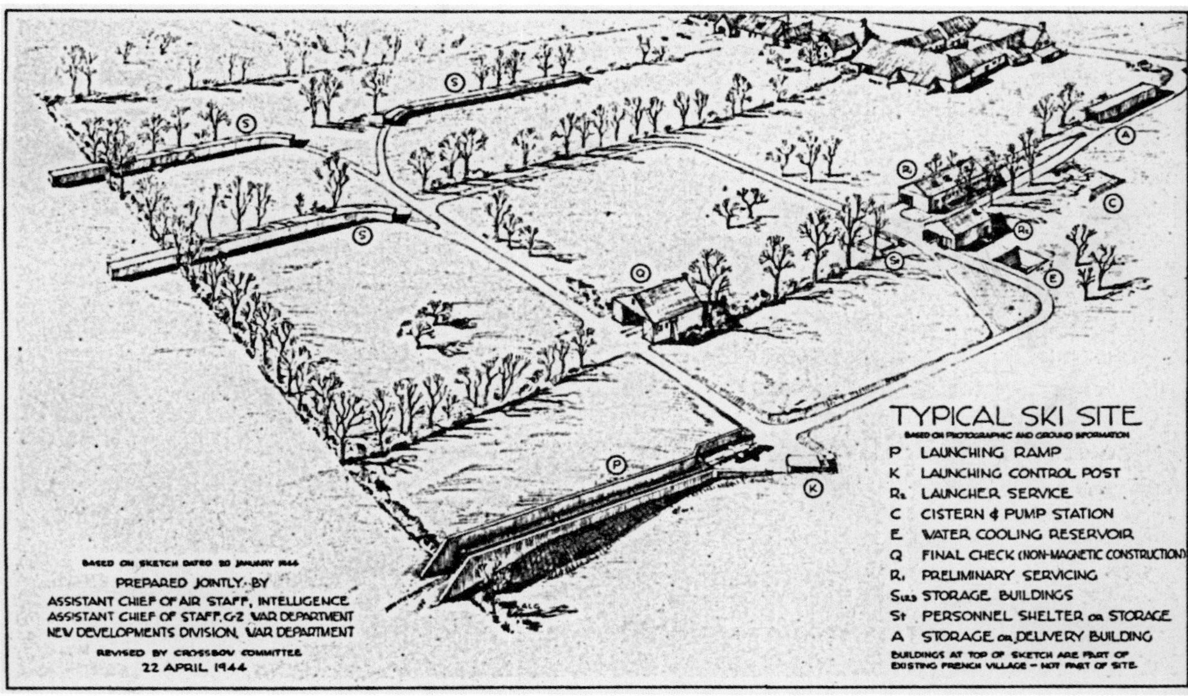

Photograph showing the V1 ramp and a reconstructed flying bomb being launched.
(Museum Blockhaus at Éperlecques)

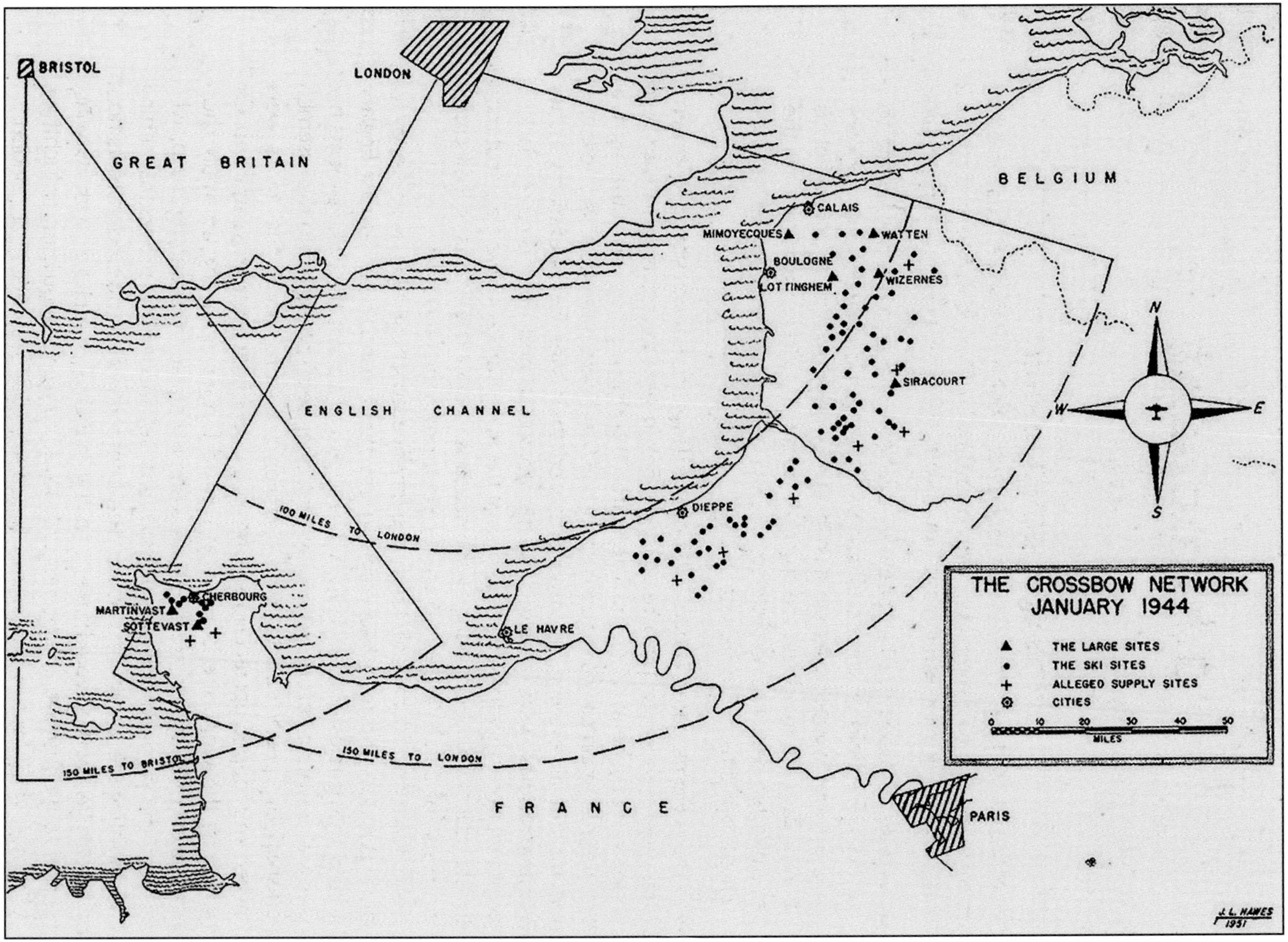

A sketch showing the various V1 and V2 launch sites. *(Cross Bow Committee)*

Chapter Four

Launching

'Target 42' (London)

On Monday 12 June 1944, in the operations bunker Wachtel addressed his commanders and told them that after months of waiting, the time had come to open fire. 'Today your waiting and your work will have their reward', he said,

> the order to open fire has been issued. Now that our enemy is trying to secure a foothold on the continent, we approach our task supremely confident in our weapons: as we launch them, today and in the future, let us always bear in mind the destruction and the suffering wrought by the enemy's terror bombing. Soldiers! Führer and Fatherland look to us, they expect our crusade to be victorious ...

With these words, commanders left for their designated launch sites departments. Orders were issued to prepare to open fire in two salvos, timed to arrive in London on 12 June at 11.40am. Thereafter, throughout the rest of the night, all sites were to maintain continuous firing for five hours, until dusk the following morning. But all evening as crews prepared their sites for operational duties; worrying reports were received of the confusion and the complete lack of equipment necessary to launch the bulk of the flying bombs. It was clear that most sites stood little chance of firing by the deadline, despite the efforts of their crews. As a result of this, Wachtel reluctantly postponed the launching.

By 3.30am on 13 June, only ten of the fifty-five launching ramps were ready for action. But now the time had come to use them in retaliation against England for its bombing raids on Germany. After months of training the Flakregiments had made the pre-launch procedure ready. First, after being brought into the assembly workshop by special vehicles with a trailer, the flying bomb fuel tanks were checked and topped up. Its tailplane, fin and rudder units were then attached. On completing the work, the bomb was moved further along the assembly shop, being taken over by the steering unit assembly squad or Steurungstrupp, who mounted the gyrocompass mechanisms and set them on a provisional setting, whilst hoisting the bomb on a pulley. The bomb was then lowered from its hoist onto a trolley, which would take it to the launching ramp. Still with its wings clamped to the fuselage, the bomb was moved to a special building or Richthaus and aligned with the launching ramp, which was pointed directly at London.

The wings were then fitted and the bomb was programmed according to range and the bearing of its target. The Navigation Group calculated its actual course and height which it would fly. With the bomb ready for firing, the launching ramp squad or Geschutztrupp moved the V-1 from the conveyor trolley and placed it slowly onto the ramp. The bomb was then attached to the ramps firing position on the underside of the fuselage. Sitting ready for take-off, the launching crew took cover inside the control 'bunker 1 or 'Kommando stand'. Once the all clear was given, the firing officer gave the order, 'fire!'. On a small portable control box, a lever was pulled, and the flying bomb's jet engine rumbled into life with a pulsating roar. A burst of hot exhaust shot out of the rear of the engine and the bomb was catapulted up the steel ramp at a speed of nearly 300mph. The V-1 then climbed into the sky, increasing speed as it gained height. Its distinctive pulsating drone could be clearly heard, something Londoners would soon come to dread.

Nine more of the bombs were launched in short succession, but five of the shots failed, falling to the ground near to the sites and bursting on impact. That morning six of the ten bombs went off as planned. The Flak crews, many of them as young as 18-years-old, looked on as the pilotless machines carrying 2,000 lbs of high explosives, headed out into the distance for England.

On its journey, one of the five successfully launched V-1s probably crashed into the Channel. The remaining bombs continued speeding through the sky on their north-north-westerly course toward the Kent coast, bound for London. As they did so, on the southern coast of England, members of the Royal Observer Corps who had been undertaking aerial reconnaissance duties from their watchtowers, suddenly identified what seemed to a 'black shape with sheets of flame spurting out of it'. The Observer Corps had been briefed to anticipate and recognise the pilotless aircraft, and relayed the code-word 'Driver'. As they set about plotting the path, the jet-propelled bombs roared on over the Kent and Sussex countryside.

The aiming point for the V-1 assault was London's Tower Bridge, and it would only take 22 minutes to reach its target, if everything went according to plan. The first bomb to cross the coast tripped its diving controls above Swanscombe near Gravesend. First, its engine cut-out, and then the bomb veered earthwards, whistling through the air at a steep angle until it hit the ground with a massive bang. The second bomb came down two minutes later on a farm in Cuckfield, Sussex, causing minor damage and killing some animals. The third bomb exploded at Crouch, near Sevenoaks and the fourth hit and destroyed a railway bridge on Grove Road in Bethnal Green, killing six people and injuring twenty-six.

For two days, London and southern England waited in anticipation for more of Hitler's 'retaliation weapons'. By 15 June, after various faults had been rectified, permission was granted to prepare to reopen their offensive against 'Target 42' (London). With all fifty-five launching sites now operational, Flak crews could achieve their plans of destroying 'in the name of the Führer'.

In the presence of Heinemann, Watchel issued his orders for the great assault on England. At 11.18pm, north and south of the Somme, hidden in woods and copses, V-1s could be seen being simultaneously launched from their ramps bound for England. For twelve hours, until the middle of the afternoon of 16 June 1944, 244 bombs were launched from fifty-five sites; forty-five of them crashed soon after take-off. But German reconnaissance aircraft brought reports of large fires in the target area. The success of the attack restored the spirits of the German High Command. Later that day, a communiqué used by OKW announced to the German public that the bombardment of southern England was successful.

For the inhabitants of southern England, the story was completely different. During the course of the V-1 attacks, seventy-three bombs fell in Greater London, the majority coming down in boroughs south of the Thames. Anti-aircraft defences managed to shoot eleven of these down, but altogether 144 bombs crossed over the English coast. Of these, seventy-one crashed in Kent or Sussex, most of them probably developing mechanical faults, or simply running out of fuel.

In the weeks that followed, the battle of the flying bomb was fought. Those living in Kent and Sussex, what came known as 'Doodlebug Alley', were to endure an ordeal comparable with London. Within three weeks of the first attack, the number of civilians killed rose to 1,769. By mid-July, the death toll was estimated at about 3,600; by the end of August, 5,476 people had been killed.

As the bombardment against England continued, conditions for the launching crews worsened under the harassment of aerial bombing. Coupled with the strains of launching the bombs and the ever-increasing worries of approaching Allied ground forces, many sites in France began to dismantle equipment, returning it to the railheads to Belgium, Holland, and Germany. From the Low Countries, Heinemann's Flakregiments intended to establish the sites for firing principally against the onrushing Allied armies. During the coming weeks, Brussels, Antwerp, and Rotterdam would endure the threat of the flying bomb.

In August, demolition groups visited all the remaining operational sites in France and laid a network of demolition charges. The last V-1 launched from France left one of the Pas de Calais ramps at 4.00am on 1 September 1944. For over two months, 2,224 V-1s had been launched. Now, for the first time, the skies above southern England fell silent to the relentless sound of the flying bomb. The battle over southern England appeared to be over. So optimistic were the British government of their victory over Hitler's vengeance weapons, they began issuing orders to suspend future counter offensives against V-2 rockets 'as it no longer presented a possible danger' – they were wrong.

By the end of August 1944, whilst the British government were congratulating themselves, a plan had been secretly drawn up for the launches against London and Paris. This time it would not be the V-1 fired in anger, but a greater threat, the supersonic V-2 rocket. At first, the proposed launch area was to be from Belgium, between Tournai and Ghent. But because of the speed of the advancing Allies and the capture of Brussels on 3 September, the firing zone was subsequently moved further north to The Hague in Holland. Altogether there were two launch groups, Gruppe Nord to attack London and Gruppe Sud for attacks against Belgium and France.

For the V-2, there were to be no fixed launching sites as previously planned. Instead, they would use mobile launchers specially converted to be transported across country. Everything, including its supporting crew, equipment, and fuel, was designed with mobility in mind. In transit, the V-2 was laid out at full length on a Meillerwagen trailer in its special camouflage protective cover. The Meillerwagen was so versatile that it not only transported the rocket to site, but also served as a firing platform. Because of this unique design, almost any surface could be used to launch the rocket, even a road or a clearing in a forest. Once the launching units had arrived at their designated area to prepare the rocket for launching, firing crews and engineers activated the Meillerwagen hydraulic lifting pistons, which slowly raised the missile to its vertical position. This took about fifteen minutes.

Once its nose was pointing straight up, locked in its firing position, a mass of electrical cables were then connected, and tests on various circuits and systems were carried out. Engineers then rolled up on wheels the lightweight aluminium liquid oxygen and alcohol tanks to begin fuelling the 46-foot rocket. The Meillerwagen served as a gantry tower during the fuelling process. Next, technicians performed a number of operations including activating the rocket's gyroscope mechanism. In all, it took just over an hour to prepare the V-2 for launch. After one final check to make sure it was level, the launching crew installed the rocket's electronic ignition.

As the final countdown for liftoff began, the Meillerwagen was moved away from the missile, and except for the unit commander and a few engineers and technicians, who were already in the trailer, everyone hastily took cover in a freshly dug trench. As the last seconds ticked away, inside the trailer, the commander peered through the narrow observation slits and shouted out the order 'Ignition!'. Sitting near the officer, the propulsion engineer pushed the first of three buttons on his panel of switches and gauges. Whilst everyone else looked on at the streams of vapour liquid pouring out of the rocket's vents, the V-2s tail suddenly ignited with its initial thrust of 8 tons lifting off the ground. After 23 seconds, the rocket entered 'sonic velocity'. Twelve seconds later, it was travelling more than 1,200mph. At altitude, it continued to increase in speed as it sped through the stratosphere moving more than 3,200 feet per second. In just over four minutes, the V-2 'burnt out' and crashed down, impacting at supersonic speed into its target area in London.

On 7 September 1944, the SS, who had now taken over control of the V-2 programme, arrived in the city of The Hague with their deadly cargo, and gave its inhabitants in the districts of Konijnenlaan, Koekoekslaan, and Lijsterlann only two hours to evacuate the area. Early the next morning, a column of trucks and launch vehicles towing Meillerwagen trailers arrived and began preparing for a double launch in Koekoekslaan.

The first V-2 offensive was fired by a demonstration unit Lehr und Versuchs Batterie 444 at Paris at 8.30am on the morning of 8 September. That evening Gruppe Nord began launching its rockets against London. At 6.40pm, with no warning, there was a huge explosion in Chiswick, 8 miles off target, where it killed three people and injured ten. Seconds later, another rocket fell harmlessly to earth near Epping in Essex. Within an hour the launching team had left Koekoekslaan, leaving no trace of the launching site.

In the weeks to come, more rockets were launched in a number of residential areas, and in the surrounding parks of the city. Various other Dutch towns were also used for launching. Just north of the town of Zwolle, for instance, units fired-off rockets against the English towns of Norwich and Ipswich. In November, the number of V-2s launched against England increased dramatically: eighty-two rockets landed in the streets of London, killing hundreds of civilians. The worst incident was at New Cross Road when a V-2 fell on Woolworths, killing and burning to death 160 people. Continental targets also suffered equally from the V-1 and V-2 attacks. Antwerp alone had been bombarded by 924 rockets up to the end of 1944, compared to 447 directed at London. Hundreds of Dutch civilians also died from defective rockets that veered off course and crash-landed.

The last rockets of the war were fired from The Hague during the early evening of 27 March 1945. The last V-2 exploded in London, killing 134 people and maiming and injuring forty-nine others in a block of flats in Whitechapel. In total, the rocket attacks had killed 2,754 people and injured 6,523.

A V-2 rocket can be seen being prepared for launching. These rocket gantries were mobile and weighed 15 tonnes. These gantry cranes were used by Wehrmacht maintenance units for tasks such as lifting tank turrets, and also by V-2 crews.

Meillerwagen trailer was so versatile that it not only transported the rocket to site, but also served as a firing platform. Because of this unique design almost any surface could be used to launch the rocket, even a road or a clearing in a forest. Once the launching units had arrived at their designated area to prepare the rocket for launching, firing crews and engineers activated the Meillerwagen hydraulic lifting pistons which slowly raised the missile to its vertical position. This took about fifteen minutes.

Also part of the launching unit facility was a utility trailer converted as diesel air compressor and fuel tanker.

A V-2 on a Meillerwagen trailer for launching.

The first of two photographs taken in sequence at Peenemünde showing a V-2 rocket being prepared for launching from a rail line. V-2 rockets were transported by train and from the railhead, they were transferred to mobile Meillerwagen trailers for transport and erection at the launch site.

Three photographs showing the launch of the V-2 rocket. Even though the V-2 installations were bombed, the V-2 was practical and versatile enough to be launched from a mobile Meillerwagen. This mobility made the launch units extremely difficult for Allied air forces to target and destroy.

In preparation for launching is a V-2. Note the three-section extendable ladder known as the Magirus ladder attached to the weapon for access.

Two photographs showing preparation for launching from the Meillerwagen trailer. After initial transport by rail, the rocket would be moved onto the Meillerwagen from a separate road transporter, the Vidalwagen, at a transfer point. The Meillerwagen would then be towed to a camouflaged launch site. Once in position, the trailer could elevate the missile to a vertical firing position for fuelling and launch preparation.

A V-2 launching unit can be seen sitting inside a halftrack prime mover, which is towing a Meillerwagen trailer complete with V-2 rocket.

This photo captures the moment a V-1 flying bomb is about to be launched from its ramp, destined for London in 1944.

A V-1 flying bomb has just been launched from its ramp, destined for southern England.

A V-1 flying bomb has hit part of central London.

A combustion chamber of a V-2 rocket that crashed in East London in March 1945. Other V-2s crashed in London without exploding and considerable data and intelligence was gained from them.

The illustration has been adapted and re-drawn by the author from an idea in Michael J. Neufeld's book *The Rocket and the Reich*. (The Smithsonian Institution)

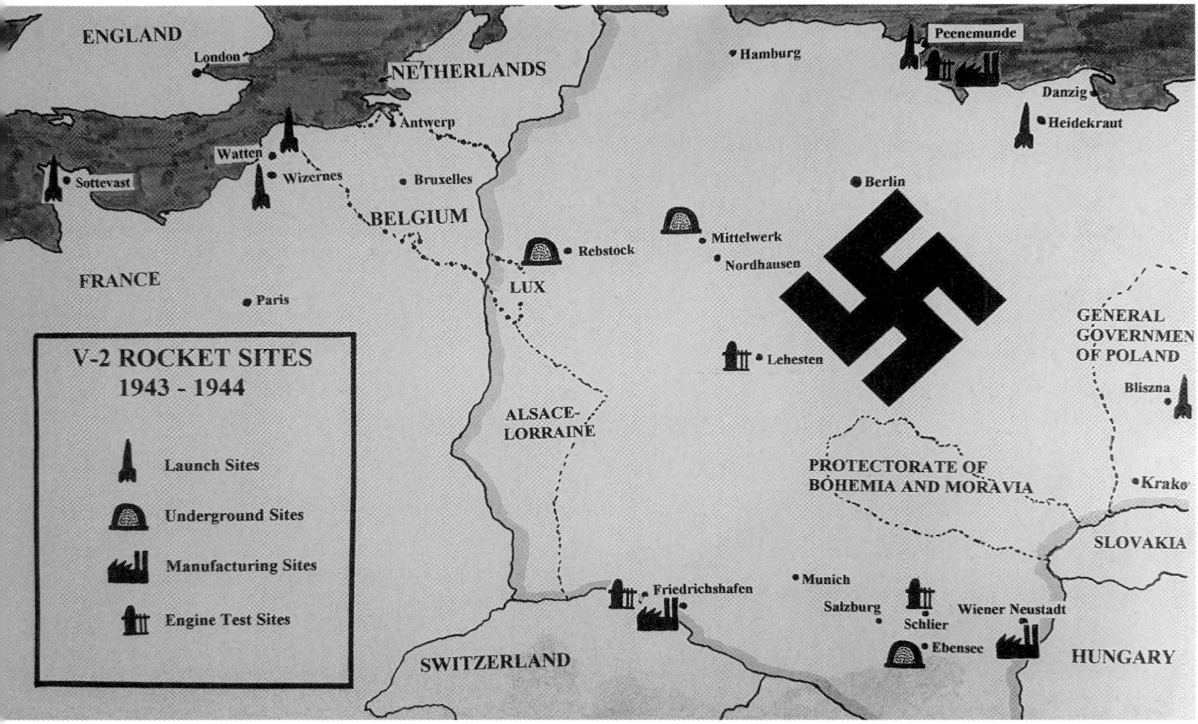

A young girl stands inside a damaged building after a V-2 exploded in Battersea, London on 27 January 1945.

Chapter Five

V-3 Cannon

Wiese (meadow) and Bauvorhaben 711 (Construction Project 711)

Whilst plans and testing were undertaken for the V-1 and V-2 rockets including blueprints drawn for their launching sites, another secret idea had been put forward as early as 1942 for a Vergeltungswaffe 3 or Vengeance Weapon 3. The idea was to design a massive, long calibre single charge gun that could be fired against London. The design of the gun was to fire a projectile down the long barrel and as the charge travelled, it would increase in speed using hot gases that propelled and accelerated the projectile. In order to keep the project secret, the V-3 was given a codename, Hochdruckpumpe or High Pressure Pump, HDP for short. It was also known as Fleißiges Lieschen or Busy Lizzie.

The long barrel was designed to have side chambers which would provide additional acceleration of the shell during its passage up the barrel of the gun. This allowed a muzzle velocity of over 1,500 metres per second. For firing at a target, the cannon was arranged in regular pairs and angled to project the projectiles thrust at its maximum velocity. The layout of the gun barrel was later nicknamed by the German designers as the Tausendfüßler or millipede.

To fire the gun at London, it was planned to use two batteries which would fire the weapon in succession using hundreds of projectiles weighing some 140 kilograms with an explosive charge of 25 kilograms.

The first test that was performed on the gun by August Coender, developer and chief engineer of the plants Rochling Stahlwek AG, was very encouraging. It immediately caught the attention of the Ministry of Arms and Coender presented the gun to Albert Speer as capable of firing massive projectiles from the coast of the Pas-de-Calais on London.

In May 1943, Speer took the proposal to Hitler. Although he received a lukewarm response from his Führer, it was not until the RAF bombed Peenemünde rocket facility on 17 August that Hitler finally agreed that the gun be built without further tests. Almost immediately, Coender was set-to-work to construct the gun, and despite problems with producing a shell that could be fired at such a speed, his team of engineers built a single-full-sized 490-foot barrel at Misdroy on the Baltic island of Wolin, near Peenemunde.

Whilst the gun was being built, an area in northeast France outside Pas-de-Calais had been chosen as a launch site. Major Bock of Festung Pioneer-Stab 27 (the fortification regiment of LVII Corps, Fifteenth Army), had been given the order of finding a suitable location for the HDP batteries. Engineers had informed Bock that in order to ensure the gun fired properly, a hill with a rock core would be most suitable, as the gun tubes could be installed into inclined tunnels. This would also include construction of adjacent tunnels for support equipment and supplies. Unlike the V-1 or V-2, the V-3 would not be able to be moved and would be permanently fixed and aimed at London.

In the spring of 1943, a site was finally selected at a limestone hill just north of the Hidrequent quarries near Mimoyecques in the Pas-de-Calais area, very near to where the V-1 and V-2 sites were already under construction. The construction site for the V-3 was code-named Wiese (meadow) and Bauvorhaben 711 (Construction Project 711), and in September 1943, Organisation Todt began work on the site. Construction commenced with the building of a railway line and followed a few weeks later excavating the gun shafts. A total of five inclined tunnels were to be built to hold five of the gun tubes, which were designed at 50 degrees and stretched 344 feet below the hilltop. The five inclined tunnels exited the hilltop through massive 98 feet wide and 18 feet thick concrete slabs. All five openings were reinforced and protected by steep plates and special armoured doors. Numerous tunnels and elevator shafts were dug by thousands of slave labourers under appalling conditions. All of these tunnels were built to support the guns. To prepare the gun for operation, some 1,000 soldiers from Artillerie Abteilung 705 and supporting units were deployed to Mimoyecques. Artillerie Abteilung 705 had been organised in January 1944 under Oberstleutnant Georg Borttscheller to operate the Wiese gun facility.

However, following extensive aerial Allied reconnaissance missions across the Pas-de-Calais area, surveillance had detected the V-3 site at Mimoyecques. On 6 July 1944, allied forces, chiefly the RAF's No. 617 Squadron carrying deadly Tallboy bombs (deep penetration earthquake bombs), targeted the underground complex, causing significant structural damage, flooding, and collapsing tunnels. As a result of the heavy bombing, it rendered the German supergun project unusable before it could be completed.

With Construction Project 711 now disbanded in northern France, the V-3 gun was never installed at the complex. As a result of the allied attack, the SS took control of the project, and SS-Obergruppenführer General Hans Kammler who was responsible for civil engineering projects, including the V-weapons programme, ordered that the gun be adapted and shortened to 160 feet long with twelve side-chambers. The SS were eager to get the gun ready for action and instructed the Wehrmacht artillery unit Artillerie Abteilung 705 under the command of Hauptmann Patzig to prepare the V-3. The two guns were placed in a wooded ravine on thirteen steep support structures complete with solid wooden bases at a 34-degree slope along the Ruwer River at Lampaden just southeast of Trier in Germany. The area was designated as Target No. 305. This included the construction of concrete blockhouses, which were erected between two gun tubes, and numerous smaller bunkers which were to store the projectiles and propellant charges.

The Germans nicknamed the V-3 as the 'Lampaden guns' and these were prepared for the action for the Ardennes Offensive, which was the last German offensive on the Western Front in December 1944. However, by the time the offensive was launched on 16 December, the gun was still not ready for action. By the end of the month, one of the guns fired off two warm-up rounds followed by five high-explosive shells. One of the two guns was then deployed at Buhl, aimed at Belfore supporting the German Operation Nordwind offensive. However, when the offensive failed, the gun was dismantled and equipment removed. The gun tube was also used against Luxembourg and fired off some 142 projectiles into the city, which killed ten and wounded thirty-five.

On 15 February 1945, one of the 'Lampaden guns' was dismantled when the US Army units advanced within 2-miles of the Lampaden site. Yet, in spite of the advance of allied forces, there were still plans to deploy batteries to bombard London, Paris and Antwerp, but this idea, due to logistical problems, was abandoned. As a result, Artillerie Abteilung 705 discarded the V-3 and was re-equipped with conventional artillery for the remaining months of the war.

Mimoyecques eastern site reconstruction drawing depicting the planned V3 weapon site.

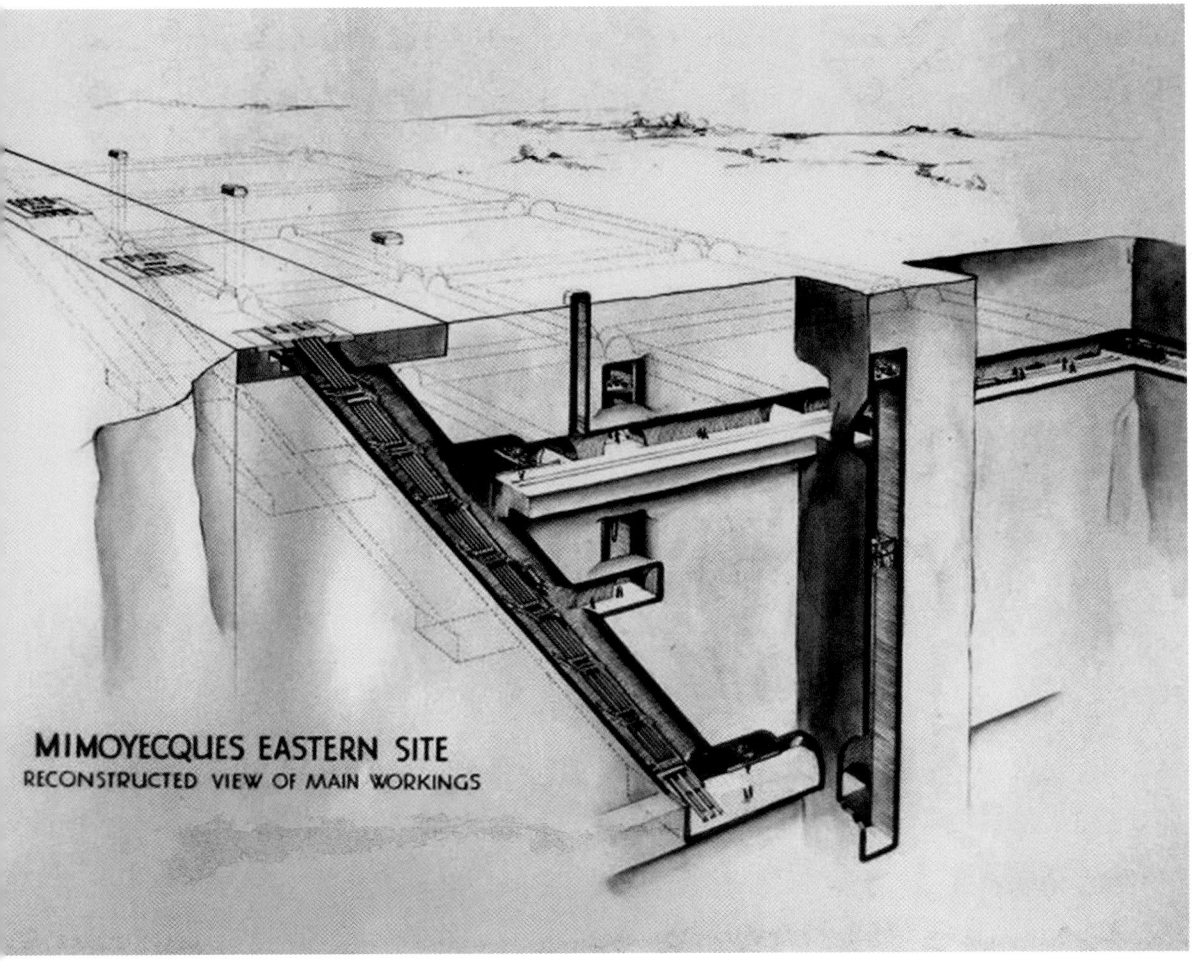

Two photographs taken in 1942 showing the trialling of a prototype V3 gun barrel at Laatzig, now in Poland. The V3 was also known as the Hochdruckpumpe (High Pressure Pump, HDF for short), which was a code name intended to hide the real purpose of the project. It was also known as Fleißiges Lieschen (Busy Lizzie). *(Bundesarchiv Bild 146-1981-147-30A / 146-1981-147-31A)*

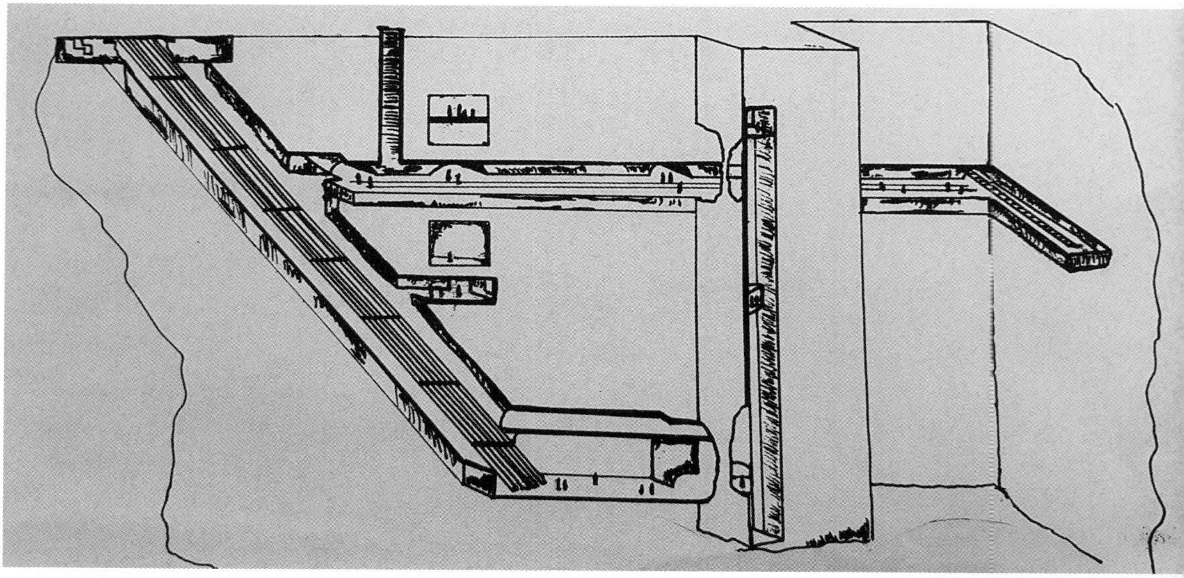

(**Opposite, top**) The author's impression of one of Hitler's most sinister secret weapons, the long-range gun site at Mimoyecques, near Calais. It was designed to fire 9 foot dart-like shells at London at a rate of 600 per hour. The underground sketch of the site shows the immense scale: note the size of the men at the foot of the 416 foot gun barrels.

(**Opposite, middle**) The V3 gun barrel under test.

(**Opposite, bottom**) The first of two photographs showing the service tunnels at the former Mimoyecques V3 rocket facility. The operational version of the gun was being constructed in the chalk cliffs of France near the town of Mimoyecques. The Germans used T-shaped sections, 4m to 5m long, assembled to a total length of 150m, which was capable of shooting shells over a 150 mile range. However, it took a large number of re-loaders to put powder in each T-arm after a shot and consequently could only be fired once every five minutes. Furthermore, every third shot caused the barrel in one of the T-sections to crack, meaning it had to be removed and replaced. Bunkers in the Pas de Calais were being built for the weapon, but they were subject to incessant bombing and finally overrun by Allied troops before they could be completed. *(Mimoyecques Museum)*

Appendix

Launch Sites

V-1 Launch Sites

FRANCE

V-1 Launch Site Réalcamp Nr. 169
Normandie
Réalcamp
Vergeltungswaffe 1

Launch Bunker Site V1 Nr. 170
Normandie
Réalcamp

Launch Ramp Site V1 Nr. 671
Normandie
Campneuseville

Calibration Plate Site V1 Nr. 170
Normandie
Réalcamp

Site V1 Nr. 180
Normandie
Réalcamp

V-1 Launch Base Avesnes-Chaussoy
Hauts-de-France
Avesnes-Chaussoy

V-1 Feuerstellung Ligescourt
Hauts-de-France
Ligescourt

V-1 Feuerstellung Fiefs
Hauts-de-France
Fiefs

V-1 Feuerstellung 622 – FA Depot
Hauts-de-France
Heuringhem

V-1 Feuerstellung 622 – Waterbasin
Hauts-de-France
Heuringhem

V-1 Feuerstellung 622 – FA Vormontage
Hauts-de-France
Heuringhem

V-1 Feuerstellung 622 – FA Dampferzeuger-Garage
Hauts-de-France
Heuringhem

V-1 Feuerstellung 620 – FST
Hauts-de-France
Moringhem

V-1 Feuerstellung 620 – Richthaus
Hauts-de-France
Moringhem

V-1 Feuerstellung 620 – FA Building
Hauts-de-France
Moringhem

V-1 Feuerstellung 620 – FA Pumpstation
Hauts-de-France
Moringhem

V1 Feuerstellung 620 – FA Stofager
Hauts-de-France
Moringhem

V-1 Feuerstellung 620 – FA Dampferzeuger-Garage
Hauts-de-France
Moringhem

V-1 Feuerstellung 620 – FA Vormontage
Hauts-de-France
Moringhem

V-1 Feuerstellung 618 – FST
Hauts-de-France
Journy

V-1 Feuerstellung 618 – Remains FA Building
Hauts-de-France
Journy

V-1 Feuerstellung 618 – FA Gebäude
Hauts-de-France
Journy

V-1 Feuerstellung 623 – FA Lager
Hauts-de-France
Morbecque

V-1 Feuerstellung 623 – Zunderlager
Hauts-de-France
Morbecque

V-1 Feuerstellung 623 – Richthaus
Hauts-de-France
Morbecque

V-1 Feuerstellung 623 – FA Abschußbunker
Hauts-de-France
Morbecque

V-1 Feuerstellung 623 – Abshußrampe
Hauts-de-France
Morbecque

V-1 Feuerstellung 623 – Waterbasin
Hauts-de-France
Morbecque

V-1 Feuerstellung 623 – FA Vormontage
Hauts-de-France
Morbecque

V-1 Feuerstellung 623 – FA Depot 3
Hauts-de-France
Morbecque

V-1 Feuerstellung 623 – FA Depot 2
Hauts-de-France
Morbecque

V-1 Feuerstellung 623 – FA Depot 1
Hauts-de-France
Morbecque

V-1 Feuerstellung 623 – Concrete Roads
Hauts-de-France
Morbecque

V-1 Feuerstellung Val Ygot
Normandie
Ardouval

V-1 Feuerstellung Yvrench
Hauts-de-France
Yvrench

V-1 Feuerstellung Siracourt (Wasserwerk 1 St.-Pol)
Hauts-de-France
Siracourt

V-1 Feuerstellung Wallon-Cappel
Hauts-de-France
Wallon-Cappel

V1 Storage Facility Renuscure
Hauts-de-France
Renuscure

NETHERLANDS

V-1 Launch Site Dortherdijk
Gelderland
Joppe

V-1 Launch Site Dortherdijk
Gelderland
Joppe

V-1 Launching Site Tusveld Bornerbroek
Overijssel
Bornerbroek

V-1 Launching Site Bornebroeksestraat Borne
Overijssel
Borne

Launch Basis V-1 Mosselenbuurt Bergen
North Holland
Bergen

V-1 Fallback Ampsen
Gelderland
Lochem

V-1 Memorial Harfsen
Gelderland
Harfsen

V-1 Launch Platform Nijreesbos Almelo
Overijssel
Almelo

Launch Basis V-1 Airport Bergen
North Holland
Bergen
Fliegerhorst Bergen

Launch Site V-1 Hellendoorn
Overijssel
Nijverdal

V-1 Feuerstellung Camping Twente
Overijssel
Enschede

V-1 Feuerstellung Haaksbergen
Overijssel
Haaksbergen

V-1 Feuerstellung Assinkbos
Overijssel
Haaksbergen

V-1 Feuerstellung Helmerhoek
Overijssel
Enschede

V-1 Feuerstellung Castricum
North Holland
Castricum

V-1 Feuerstellung Warmond
South Holland
Warmond

V-1 Feuerstellung Averlo
Overijssel
Averlo

V-1 Feuerstellung 519 Lettele
Overijssel
Lettele

GERMANY

V-1 Launch Site Rankenhohn
Nordrhein-Westfalen
Eitorf

V-1 Feuerstellung Peenemünde-West
Mecklenburg-Vorpommern
Peenemünde

V-1 Feuerstellung 328 Bruchhausen
Rheinland-Pfalz
Bruchhausen

V-1 Feuerstellung 303 Kuchem
Nordrhein-Westfalen
Kuchem

V-1 Feuerstellung Zempin
Mecklenburg-Vorpommern
Zempin

ITALY

V1 Feuerstellung Ghedi
Lombardy
Ghedi

BELGIUM

V-1 Feuerstellung Proven
West Flanders
Proven (Poperinge)

POLAND

German Bunker Blizna
Subcarpathian
Ocieka

V-2 Launch Sites

NETHERLANDS

V-2 Missile Emplacement Heek
Nordrhein-Westfalen
Heek

V-2 Launch Site Hessum
Overijssel
Hessum

V-2 Launch Site Dalfsen
Overijssel
Dalfsen

V-2 Launch Site Vrederust
Zeeland
Serooskerke

V-2 Launch Site Hoek van Holland
South Holland
Hoek van Holland

V-2 Launch Site Beukenhorst
South Holland
Wassenaar

V-2 Launch Site Raaphorst
South Holland
Wassenaar

V-2 Launch Site Statenkwartier
South Holland
Den Haag

V-2 Launch Site "Zorgvliet"
South Holland
Den Haag

V-2 Launch Site Waterpartij
South Holland
Den Haag

V-2 Launch Site Landgoed Ockenburgh
South Holland
Den Haag

V-2 Launch Site Landgoed Duindigt
South Holland
Wassenaar

V-2 Launch Site "Bloemendaal"
South Holland
Den Haag

V-2 Launch Site Naald van Rijswijk
South Holland
Den Haag

V-2 Launch Site Wassenaar
South Holland
Wassenaar

V2 Launch Site Rijsterbos
Friesland
Rijs

BELGIUM

V-2 Launching Site Petites Tailles
Liège
Petites Tailles

V2 Launch Location Gouvy
Luxembourg
Gouvy

GERMANY

V2-Werks Friedrichshafen
Baden-Württemberg
Friedrichshafen

Peenemünde
Mecklenburg-Vorpommern
Peenemünde

Launch Complex Leba
Pomeranian
Leba

FRANCE

Underground V2 storagetunnels
Hauts-de-France
Bergueneuse

V2 base La Coupole
Hauts-de-France
Helfaut

Blockhaus d'Eperlecques
Hauts-de-France
Éperlecques

POLAND

German Bunker Blizna
Subcarpathian
Ocieka

V-3 Launch Sites

FRANCE

Bauvorhaben 711
Fortress of Mimoyecques

LUXEMBOURG

Target No. 305
Lampaden

Notes

Notes